GLOBAL
DUST BOWL

GLOBAL DUST BOWL

Can We Stop the Destruction of
the Land before It's Too Late?

C. Dean Freudenberger

Augsburg ∎ Minneapolis

GLOBAL DUST BOWL
Can We Stop the Destruction of the Land before It's Too Late?

Scripture quotations unless otherwise noted are from the Revised Standard Version of the Bible, copyright © 1946, 1952, and 1971 by the Division of Christian Education of the National Council of Churches.

Cover design: Lecy Design

Library of Congress Cataloging-in-Publication Data
Freudenberger, C. Dean, 1930–
    Global dust bowl : can we stop the destruction of the land before
it's too late? / C. Dean Freudenberger.
        p.   cm.
    Bibliography: p.
    ISBN 0-8066-2448-5
    1. Sustainable agriculture.   2. Agricultural ecology.
3. Agriculture.   4. Agriculture—Moral and ethical aspects.
I. Title.
S494.5.S86F74   1990
333.76'137—dc20   89-15202
                                                            CIP

The paper used in this publication meets the minimum requirements of American National Standard for Information Sciences—Permanence of Paper for Printed Library Materials, ANSI Z329.48-1984.   ∞™

Manufactured in the U.S.A.   AF 9-2448
94   93   92   91   90   1   2   3   4   5   6   7   8   9   10

CONTENTS

PREFACE

In 1974, after 17 years of agricultural development work in Africa, Southeast Asia, the Caribbean, some of the Pacific Islands, and parts of Latin America, I was asked by an editor of *The Los Angeles Times* to write a 2,500-word article about the causes of world hunger and what can be done about it. I turned in a 20-page paper, but the response to my hard work was: "We need something simple." At that moment I discovered the need for greater public understanding of a most urgent problem: hunger and agriculture.

In a society so thoroughly urbanized and industrialized, it is easy to forget that agriculture is fundamental for human life. In our time agriculture is reduced to the production of crops for markets. This reductionism has resulted in erosion of rural life and the destruction of natural resources. The agricultural crisis, with its growing threat to the health and future of the planet, is a crisis of culture. It is a crisis in human self-understanding and responsibility.

Hunger, rural poverty, and environmental destruction is the legacy of nearly 400 years of colonialism. Agriculture in the colonies was designed for every conceivable purpose except the self-reliance of indigenous peoples and care for the environment. Only recently have these issues been discussed. Foreign policy for international agricultural development has yet to account for environmental needs, let alone to make a transition from export crop agriculture to regenerative, domestic, self-reliant food systems. An agricultural ethic has yet to be found.

In response to the present need, this book proposes that our present system of industrial agriculture be replaced by agroecology (regenerative agriculture that preserves and enhances natural resources). In order to make the transition from industrial agriculture to agroecology, we must place long-range ecological concerns above short-range commercial ones. The question about human purpose and responsibility for the health of our planet must now be asked.

Earlier I wrote *Christian Responsibility in a Hungry World* with the help of Professor Paul Minus.[1] Next, the filmstrip series entitled *A World Hungry* was produced with the assistance of the Franciscan Communications staff of Los Angeles.[2] As the U.S. farm crisis began to unfold, I wrote "The International Impact of the U.S. Farm Crisis" for *Farming the Lord's Land,* edited by Charles Lutz.[3] In 1984, *Food for Tomorrow?* was written to describe the magnitude of the farm crisis.[4] This book, in part, was based on the findings of the American Farmland Studies developed during the Carter administration under the direction of Secretary of Agriculture, Bob Bergland.[5] The thesis of *Food for Tomorrow?* is that if we continue on our present course in U.S. agriculture, there will not be food for tomorrow.

In *Food for Tomorrow?* I stated:

Facts like these reveal the magnitude and the urgency of the crisis in world agriculture, a crisis affecting all the people of the entire globe. This crisis is more radical and more serious than any other we face—except perhaps the possibility of nuclear annihilation. Radical change in our relationship to the land, to the world ecosystem, is urgent. And change is possible. In order to make this change a reality, we must first change our minds about the land and our relationship to it. We need to understand the fragile nature of our global system and the magnitude of the stress we have placed on it.[6]

The thesis of *Food for Tomorrow?* is as urgent today as when it was first argued. At the School of Theology at Claremont, California, I teach in the areas of environmental ethics, rural ministry, and ecumenical social thought. At California State Polytechnic University I offer a course each year in either tropical agricultural ecology or approaches to rural community and agricultural development in the tropical and semitropical world. In addition, I serve on the animal welfare committee of this university as required by the National Institutes of Health. I also serve with the design team of the School of Landscape Architecture as it works for the development of the Institute for Regenerative Studies, an interdisciplinary effort involving the schools of architecture, environmental design agriculture, and the humanities.

I watch two of my sons, each funded by Fulbright scholarships (working in Senegal and Australia), move ahead with the development of their insights and skills in public policy, integrated agro-forestry in Sahelian rehabilitation, animal and plant community biology and ecology. One of my daughters pursues graduate studies in theology and ethics in the context of the wider question of the ecology of human existence. My other daughter and her husband from their perspective as citizens of Australia, watch the battle to preserve the integrity of the Tasmanian Forest which is presently under the pressure of the hardwood timber industry and the growing cattle and sheep stations in the "outback." My wife works with family members as a professional librarian, doing bibliographical research in order to keep us abreast of current literature.

From this context, I envision more promising futures for agriculture and the global food system with which our own nation is so intertwined. We must develop sustainable and regenerative food systems that are both effective and humane. It may be possible not only to rehabilitate the land but to enhance it.

ACKNOWLEDGMENTS

I wish to express my gratitude for those who made this book possible. In particular I thank the members of the Board of Trustees of the School of Theology at Claremont, California, for granting me a sabbatical research leave so that the first draft of this book could be written. A hearty thanks goes to Roland Seboldt, retired director of Book Development for Augsburg Publishing House, who encouraged me to record my reflections about agriculture and its future. To Dr. Irene Getz, my editor at Augsburg, I extend my thanks for her help and encouragement. Appreciation also goes to Nancy Koester, who provided many long hours of skillful service in refining the manuscript. As this book took shape, my wife, Elsie, served our team by providing her professional skills in bibliographical development. A final expression of thanks goes to you readers who give your time and energy to work through this material so that together, with concern and imagination, we can stop the destruction of the land before it is too late.

C. DEAN FREUDENBERGER

1

Some Fundamental Questions

Why Care for the Earth?

In the village of Skibet in Jutland, Denmark, there is a beautifully landscaped graveyard adjacent to the community church. At least a third of the gravestones there bear the inscription *tak*. The word *tak* means "thanks." More often than not, a small sparrow carved from granite is perched on the edge of the stone as a symbol of gentleness, beauty, and care. This simple gratitude is perhaps the best answer to the question, Why care about the earth?

In the Danish tradition the word *tak* includes no hint of fear or hope of reward for good living. Simply, *tak* expresses thanks for life and our moment of participation in it. One can observe this ethos of gratitude expressed across the Danish countryside, rural towns, barnyards, dairy parlors, and swine units—grateful care and gentle use.

Today the basic question for Western culture is, Why care for the earth? As will be examined in a later chapter, the historical record of earth caring is not good at all. In almost every generation a few lone prophets have called us to care for the earth, but their words have not been heeded. Since there is little precedent to follow, we hardly know how to respond to the question, Why Care?

I began to find an answer to this question during the ten years that I lived in Africa. I came to realize that everything centered in caring. One's identity, self-esteem, and position within the community depended upon how well one respected the memory and wisdom of the ancestors, how carefully one used the land for the good of the present and future communities that would depend upon it. These cultural insights were the beginning of my quest for earth caring, which has led me also to the Danish village and the North American farm.

A responsible society expresses its gratitude in the free choice to care. True freedom is not freedom from responsibility but freedom for responsibility. To choose to be responsible and to care about the integrity of things is an expression of gratitude. Such a choice is not, in mature ethical behavior, based on the possibility of rewards or punishments, like a child who shares a toy in the hope of some reward. Both Danish and African villages helped me learn that true freedom lies in grateful responsibility.

When economic and technological structures make it impossible for a person to choose to be responsible, a basic human right is denied. I shall never forget the time I was having a cup of coffee one morning in a farmhouse in the upper Midwest, discussing the needs for new terraces, grassed waterways, and a new windbreak. Tears filled the farmer's eyes. He said, "Because of the financial crunch in U.S. farming, I have not been able to invest in my land for the past five years." The crisis in U.S. agriculture today denies a person a basic human right: to act responsibly and in gratitude in order to care for the land. Something is wrong with the system!

If one chooses to care for the land in freedom and gratitude, one is not free to abuse the earth. Rather, justice and environmental integrity influence one's actions. One must ask, What is the responsible thing to do in order to maintain the land's integrity now and in the future? Our freedom or abuse of freedom in relation to the land has consequences; for if I choose not to be responsible, things may go well for a while but in time they will start to go bad. The ecology of the earth points to the necessity of cooperation, whether one chooses to cooperate or not.

Do We Have Any Precedents for Caring?

Donald J. Hughes, professor of classical studies at the University of Denver, wrote a little book entitled *Ecology in Ancient Civilizations*.[1] He shows that in classical times there were voices that called for the necessity of caring for the land. Plato, for example, warned of the consequences of the destruction of the pine and oak forests that covered the islands of the Greek world. But, like many others that followed him, Plato was ignored. The need for ships for the Greek navy and merchant marine prevailed. Consequently the forests were exhausted and the springs dried up, leaving only enough water for bumblebees. Today, second to the devastated African Sahel, the Mediterranean basin is the most environmentally stressed and resource-empoverished area in the world.

Professor Hughes ended his study with the observation that throughout the ages the prophetic voices have been ignored. He left his readers with the question, What makes us think there is hope in our time that prophetic voices about similar resource and environmental issues will be heard?[2]

This question about hope, like the question of caring, has to be asked. What gives us hope that the many stern voices of our time who warn us to care for the earth will be heard and acted upon? Will we heed the warnings of Rachel Carson, Carl Sagan, Barry Commoner, John Myers, René Dubos, Aldo Leopold, W. C. Lowdermilk, Liberty Hyde Bailey, Richard St. Barbe Baker, Wes Jackson, the corporate voices of the United Nations Environmental Program, the World

Meteorological Organization, *Global 2000: Report to the President,* or the Worldwatch Institute?[3] Will the demands of business and military continue to determine our use of the land?

Can We Generate the Ethical Will to Care?

A final question is posed by Wes Jackson in *New Roots for Agriculture:*

> In the earliest writings we find that the prophet and scholar alike have lamented the loss of soils and have warned people of the consequences of their wasteful ways. It seems that we have forever talked about land stewardship and the need for a land ethic, and all the while soil destruction continues, in many places at an accelerated pace. Is it possible that we simply lack enough stretch in our ethical potential to evolve a set of values capable of promoting a sustainable agriculture?[4]

Where will the ethical will to care about the land come from?

Some people suggest that we must wait until chaos is upon us and the Messiah returns to usher in a new reign of peace and harmony.[5] We saw this attitude in the pronouncements of former Secretary of the Interior James Watt when his policies about resource use were questioned. Unfortunately such "end-time" thinking is narrowly anthropocentric; it assumes that the world is designed simply for human beings, making this planet merely the stage for the human drama. Such thinking ignores the integrity of creation itself, perpetuating the false assumption that the land belongs to us.

Others may find answers to Jackson's question in the growing discontent with a science, a technology, an industry, and an economic order that are insensitive to the environment. Perhaps our ethical potential can be galvanized by new perspectives of our spaceship earth taken from outer space.

Perhaps our ethical potential can be increased by new information about spreading deserts, the exhaustion and poisoning of our aquifers, and the dissipation of the vital shield of ozone. It may be that out of the growing stress upon global agricultural systems there will emerge creative suggestions for alternatives. Our prevailing understanding of the world must change. The earth is not an object that we control but a subject to which we stand in relation. We must change our mechanistic attitude to an ecological understanding in which we see ourselves as responsible participants in the delicate fabric of the biosphere. If we are to depart from the destructive history that Hughes laments and discover the ethical potential to which Jackson points, we must change our worldview.

Unless we reevaluate our relationship to this planet, we will not be able to resolve the agricultural and ecological crisis. Only as questions are raised do the prospects of answers become more real. Humankind could never fly before someone asked whether flight were possible. Polio, yellow fever, and smallpox would continue to ravage unchecked if no one had asked if prevention were possible. Is it possible to build an agriculture that regenerates essential resources at the same time maintaining the integrity of the natural systems?

We have a choice. We can raise these questions or let things go on and take the consequences. As Socrates said, "The unexamined life is not worth living." One could add that the unexamined life may make the lives of future generations not worth living. Ultimately, the choice depends upon whether or not we want to say "thanks."

Care for the health of the earth and its future and maintaining its delicate balances are the criteria for evaluating science and technology, particularly as they affect agriculture. Care for the earth also informs our sense of purpose and identity—how we fit into our land and culture.

The author of Psalm 104 sets human beings within the larger context of creation. He describes the relations between night and day, birds, fish, beasts of the forest, humankind, sea and mountains, rain and drought. The psalmist says,"O LORD, how manifold are thy works! In wisdom thou has made them all" (104:24). All creatures look to God for life, all creatures are interdependent, and the whole of creation derives its life from God: "When thou takest away their breath, they die and return to their dust. When thou sendest forth thy Spirit, they are created; and thou renewest the face of the ground" (104:29-30). These words are not poetic fancy of a premodern mind. Rather, they express profound wisdom; we are not separated from the land, but are part of it. Together with the rest of creation, we depend on God.

Why Is Agriculture of Fundamental Importance?

The reader might ask, Why the focus on agriculture? There are two answers to this question. First, agriculture is the most important activity of every society, not simply our own. We are all interrelated. Today more than half of all food consumed in the United States comes from other nations. Agriculture is the basis of the survival and security of every nation.[6] As agriculture goes, so goes the nation. For example, the demand for cheap hamburger in the U.S. was partially met in Central America. To provide grazing land, tropical forests (which had been the breeding area for migratory birds) were felled. This ended the annual migration of some birds to the cornfields of Iowa. Now, because these birds have disappeared, more pesticides have to be used in Iowa. Toxic residues have made some rural water resources in Iowa unfit to use. Meanwhile, in Central America, the peasants must live on steeper slopes so that better land can be used for grazing cattle. Adjacent to grazing lands and in rotation with them is cotton. DDT is sometimes still used for insect control in cotton fields in this humid region of the world. Since it is virtually impossible to inspect all imported meat, DDT creeps back into our own society.

As agriculture goes, so goes society. But today, in our own highly sophisticated nation, we easily forget that agriculture means food, clothing, and shelter. Our health and the health of future generations is at stake. Ultimately it is upon the health of the land that security rests not only for today, but for generations to come. If this simple insight is lost, so is our national future and the health of the planet itself. If we fail in our responsibility toward the planet and its future, we cannot claim to be civilized. The barren landscapes at the edge of famine-relief refugee camps around the world teach a grim lesson, which we ignore at our peril. Now that our species numbers more than five billion, we must grapple for the first time with the question of biospheric survival.

A second reason that agriculture is of fundamental importance can be stated this way. Agriculture is complex. It is delicate. Yet strange as it may seem, this complex and fragile activity is the most neglected and least understood work within modern society. It rests upon the shoulders of a dispersed, isolated, and misunderstood sector of our society, the farmer and the farming community. It has come as something of a revelation that farms and rural communities are interdependent. Programs of farm assistance during recent years, particularly those involving agricultural commodity price support programs, do not work any better than a weekly allowance for a child compensates for parental neglect and abuse. Social responsibility for agriculture cannot be fulfilled with patchwork forms of assistance.

This observation is reflected in the 1984–89 study conducted by the Board on Agriculture of the National Academy of Sciences entitled *Alternative Agriculture*. This report, issued on September 7, 1989, was paid for by the Department of Agriculture, the Kellogg Foundation, and four other groups. The information was gathered by a team of our nation's preeminent scientists and gives perhaps the most important confirmation of the success of agricultural practices that use biological interactions instead of chemicals. The report points to the fact that farming methods that play down chemicals have been invented and developed by farmers over the last two decades almost entirely outside the Department of Agriculture, agricultural universities, and other institutions in American farming. The academy's Board on Agriculture said it was seeking to reverse federal policies that for more than four decades have been focused on increasing the productivity of crop and livestock farms principally through heavy use of pesticides, drugs, and synthetic fertilizers.[7]

Are the Old Assumptions Still Useful?

Basic assumptions about Western agriculture are seldom questioned. But they need to be questioned if solutions are to be found. After speaking about this in Des Moines a few years ago, I overheard this conversation at a telephone booth: "There's someone here saying that it's not our fault. Come on over and listen." Apparently someone had heard a liberating word. Several people began to realize that being on the edge of failure was not necessarily a result of poor judgment or management. So much is beyond the control of the farm family, and many of the basic assumptions upon which agriculture in the northern hemisphere is built are shortsighted and inherently flawed. For example, we assume petroleum-based fertilizers, pesticides, and fossil fuels will last indefinitely. But how long will these basic resources be in abundant and relatively cheap supply? For how much longer can we put from three to ten or even twenty fossil fuel–derived calories of energy into the production of one or two calories of food energy?[8] We must ask whether our food production systems can be sustained over long periods of time. During the past 140 years, humanity has burned approximately two billion years of fossilized carbon accumulations.

Today we assume that agriculture produces food efficiently. But are the loss of topsoil, decline in soil fertility, species extinction, and depletion or contamination of groundwater "efficient"? What is the price of such efficiency? How much longer will the "bottom line" as measured in crop yields keep us from recognizing failure, neglect, and loss in our agricultural system? Does our agricultural system abuse the environment? We assume that agricultural export production is a good thing. But how does this affect trade deficits and the capacity of nations to feed themselves?

We also assume that we have unlimited land as we convert from 12 to 20 square miles of prime agricultural land every day into urban and industrial development. Is this sound social procedure? What about long-range considerations for national food security? How does our monocropping system affect nature's diversity of species of plant and animal life? Today, worldwide, one species per day moves toward extinction. Twenty years ago the rate was about one per month. In prehistoric times the rate was about one species every 500 to 1000 years. The questions must be asked, To what degree should environmental issues shape the economy and the social order? How can resource use and environmental stress be calculated into the gross national product? In modern social and economic theory, neither capitalism nor socialism seriously addresses these questions. Gunner Myrdal, Nobel Prize winner in economics (1970) raised this question long ago. Yet it remains unheeded.[9]

We assume that the dictates of corporations count for more than the judgments of the farmer long established on the land. To what extent can we rely on cost-benefit analysis and the judgment of computer programs to substitute for the skill, wisdom, sensitivity, flexibility, memory, affection, and imagination of the farmer on the homestead that has been maintained for generations? Is the genetic development of a "super cow," for example, a responsible way to manage natural systems? Does a super cow contribute to welfare and integrity of the rural community or does it mean that fewer dairy farmers are needed? We must ask those working in the field of biotechnology and genetic engineering what the purpose of their work is. Is it to improve agriculture and the rural community, or merely to generate wealth? We risk our future (as some nations have already done) unless we ask the questions that, out of fear or ignorance, we have thus far not dared to voice.

Where Do We Find Hope?

This book is about hope. Each person and every society has imaginative potential that blossoms and grows. Agricultural science and technology can contribute to rural society. Caring relationships can still be made between humanity and the land. Hope is born when we ask questions about things that trouble us most. Asking questions sets the stage for new futures that, in light of the present state of affairs in our agriculture, are so desperately needed.

Regenerative futures can be invented. In several places significant models already exist. Before we can use the insights of these models, however, we must recognize that, far from being independent of the land, we are dependent upon it. The spectacle of desertification confronts us with famine and death; the consequences of, among other things, irresponsible agriculture. But in an agriculture based on respect for and dependence on the land, we see the promise of land and community rehabilitation, preservation, and, perhaps, even enhancement.

Today good agriculture is commonly identified with high yields and maximum efficiency. Economic gain is the goal. However no one talks about "net gain." This is strange indeed. The maximization of yields is considered good. But this concept is narrowly defined and narrowly conceived. Soil fertility and soil loss, toxic residue accumulations, and nutritional quality are not part of the equation; nor is the well-being of the farmer and the rural community. Future generations are not considered. The present criteria for evaluating and practicing agriculture must be broadened.

2

Caring for the Earth: Our Noble Obligation

Trail Bikes and Squirrels

On the west slope of Mount San Jacinto, of the San Jacinto range that stretches from the San Bernardino Mountains south to the Mexican border, my wife and I built a one-room cottage with an adjoining workshop. It is a place where we can catch our breath during the course of busy years. Situated on a steep, sloping mountainside, just above a narrow unpaved road, our cottage is surrounded by pine, cedar, and oak trees.

Three years ago a family with two boys between the ages of 10 and 15 moved into a house nearby. The boys incessantly rode their heavily cleated trail bikes up and down the mile-long unpaved road. They were not allowed to ride in the village since the bikes had no mufflers and produced a terrible roar. We hoped that the boys would soon tire of their limited track and go on to other interests. But this was not the case. They were endlessly fascinated with their "machines." Soon our lives, and the life around us, unraveled. The dust from the dry summer road was bad enough. But the trails that they made off the road into the steep landscape became small gullies.

The blue jays, squirrels, and woodpeckers, and other species of birds were driven off by the noise. With these creatures went the means of forest protection. The small birds were not around to feed on the insects and fungi. The squirrels disappeared. By constantly gathering and burying acorns, squirrels assure reproduction of oak trees which, because of massive root structures, play a vital part in holding the soil on the steep mountainside. After a good snow melt and summer rainfall, the undiscovered acorns of the winter supply germinate. When, after a 200- to 300-year lifespan, an old oak tree finally topples, it is replaced by a young tree.

I finally met with the boys. I explained what was happening and suggested that they stop riding or at least slow down and install mufflers on their bikes. Their reply was, "Why should we care about blue jays, squirrels and oak trees?" Their next comment was, "This is a free country and we can do as we please." This, of course, is the self-righteous argument of those who consider only their own "rights," not the rights of others and certainly not the integrity of the land. I spoke with the parents, explaining the issue of environmental impact upon this very fragile alpine ecosystem. The response of the parents was: "Boys will be boys." Finally we had to ask the sheriff to come out to settle the matter, which he promptly did. But certainly this was a last resort, and as such it was neither creative nor neighborly. This experience has haunted me. "Why should we care?" the boys asked. How has such insensitivity come about? Can society survive if it does not care for the land? If personal whims are the measure or standard of value, how can we hope for the future?

Why We Don't Care

The roots of our Western culture, which now permeate every nation of the world, are firmly established in the history of the ancient classical world, whose traditional resources for earth caring are ambiguous. Other cultures, thanks to Western expansion and colonialism, have witnessed a breakdown in their customs and traditions about caring for the land.

The 2,000 years of Christian history have contributed greatly to Western culture. Within the Christian traditions one may find positive and negative attitudes about caring for the earth. During the first and second century of our modern era, gnosticism was the chief rival of Christianity. Gnosticism taught that we are saved from a corrupted earth by special knowledge. Origen (185–254), teaching at Alexandria, referred to the world as a "dungeon-like place." For Origen salvation meant that humans would become divine, leaving behind the corruptions of earth. These ideas were consistent with Greek thought of the first century and before. Origen, an Egyptian, like Greek thinkers who preceded him, believed in a hierarchy of being, a dichotomy of spirit and flesh, of humanity and nature. Matter would be negated in the end. In contrast, Irenaeus, bishop of Lyons, taught that nature (the earth) is God's good creation, of which humanity was a part. Though the earth has been corrupted by human sin, everything moves toward divine fulfillment. God continues creating, working for the redemption of the earth and its people. Salvation for Irenaeus meant that because Christ has freed us from sin and death, we can become fully human.

Augustine (354–430) acknowledged the beauty of creation as a manifestation of the glory of God. Beauty suggested that the creation is blessed by God. But there was a hierarchy of being that encouraged monastic asceticism as the way to a holy life.

Thomas Aquinas (1225–1274) taught that the creation of the world was an act of divine love, and our highest good is the enjoyment of God. He was an early proponent of the medieval system of God's grace and human merit which, though its goal was salvation, did not disdain the things of this world.

Bonaventure (1221–1274) taught that one must rise above this world to experience union with God. Nature or creation must be left behind; it is of lesser importance than spiritual things.

Francis of Assisi (1182–1226) taught that all of life is sacred—human beings stand in relation not only to each other, but to brother sun, sister moon, the birds, and the foxes.

In the Reformation period, Luther (1483–1546) and Calvin (1509–1564) were concerned with the human relationship to God. Their emphasis was christocentric, but in Luther's doctrine of vocation there is a great deal about the human role in caring for the creation, especially serving the neighbor. Luther's commentaries on Genesis and the Psalms also convey the reformer's joy in God's creation and are a source of strength and hope in caring for the land.

By the time of Galileo (1564–1642), Descartes (1596–1650), and Newton (1642–1727) and the beginnings of the age of Enlightenment, nature gradually became secularized. In other words, nature was not God's good creation within which or in spite of which we live the Christian life. Rather, nature was or became an object to be used. In the Enlightenment, man (and the masculine language is ironically appropriate) was "the measure of all things." It is indeed ironic that the "Age of Reason" that supposedly liberated the West from superstition and religion also saw the development of the slave trade and the rise of colonialism, with their devastating consequences for agriculture and for human relations. Animals, plants, and all of nature (including non-European humans) were simply of instrumental value. Immanuel Kant claimed that humanity is unique in that it transcends the deterministic sphere of nature. If people are independent of nature, they may use nature in any way they please.

This human centeredness was and is today strengthened by technology. The motto of the 1933 Chicago World's Fair was "Science Explores, Technology Executes, and Man Conforms." Nature was not even worthy of mention. The economic requirements of both capitalism and socialism paved the way for the widespread acceptance of the mechanistic view of nature. Only since *glasnost* have Soviet newspapers been allowed to report on the environmental problems in the USSR. Until very recently reporting on environmental issues would most likely have been condemned as anti-Soviet agitation.

Meanwhile, theology did not do much better than the secular culture. Rudolf Bultmann (1884–1976), like Kant, understood nature as a self-enclosed world. Emil Brunner (1889–1966) understood nature as nothing more than the scenery within which the history of humankind takes place. Karl Barth (1886–1968) taught that God was the "wholly other" whom we know only by revelation—certainly not in nature. Nature is simply the backdrop of human history.

Until very recently, the Scriptures of the Old Testament and New Testament have seldom been read from an ecological point of view.[1] The task of interpreting the Bible from an ecological point of view is something quite new. Only a rare voice, such as Gustaf Wingren's, challenged this prevailing theological disdain for the earthly.[2]

With few exceptions, our cultural preoccupations have excluded, perhaps even negated, care for the earth. But now that the ecological crisis threatens the very future of our planet, we must broaden our vision. This is happening, among other places, in a new kind of biblical and theological study that is ecologically aware.[3] The first steps have been taken toward nuclear disarmament and toward resolving ozone depletion. There have been worldwide consultations on desertification, fresh- and saltwater supply management, population growth, and world food supplies. A United Nations Environmental Program is in place.[4] But basic to further progress in these areas is the need to examine Western cultural history and to expand our vision to include the welfare of the earth.

Potential to Care

The young boys on their trail bikes asked, "Why should we care?" This thoughtless attitude has prevailed for too long.

Perhaps the potential to care will grow along with our aware-ness of earth's fragility and of our dependence upon this unique planet. Since the Enlightenment and continuing yet today, Western society has been enamored with technology and the scientific method. To be sure, we have reaped many benefits. Yet our society is relatively uncritical of the pur-poses, methods, and consequences of our science and tech-nology. As Reinhold Niebuhr pointed out, our "success" in technology brings not only greater benefit but greater peril and therefore greater responsibility. We must learn to eval-uate agriculture and technology not for the short-term profits they may yield but for the kind of life they sustain over a long period of time. In other words, we must consider the health of the earth that sustains us, of whose future we are a part. It is one thing to speak of "quality of life" as though this were just one more middle-class option. It is another thing to speak of the survival of the human race, which depends for its life upon the health of this planet. As someone has said, "A good planet is hard to find."

We have the potential to care. The new efforts in organic farming, the emergence of farmers' markets in cities where products are certified to be organically grown, programs such as the Land Stewardship Project across the upper Midwest, the Meadow Creek Project at Fox, Arkansas, the Land Institute at Salina, Kansas, and several interesting activities in indig-enous livestock production are simply a few illustrations of how people care for the earth.

Earth Caring

I am grateful for the work of René Dubos, particularly for his book *The Wooing of Earth: New Perspectives on Man's Use of Nature.*[5] As a Nobel and Pulitzer Prize winner, Dubos's writing was informed by a lifetime of work in microbiology and experimental pathology. His background involved many years of distinguished service as a faculty member at Rock-efeller University in New York. Dubos studied human ecology (the relationship between humanity and environment) from biological and historical perspectives. According to Dubos, humanity has a noblesse oblige (noble obligation) for maintaining the earth, both now and for the future. This sense of concern and care is what makes us human. Dubos argues that (1) all creatures have to exploit, within sustainable limits, their natural environment (their wilderness) in order to survive; (2) no creature can survive and flourish in the wilderness without shaping it to meet basic needs; (3) humanity, like all other creatures, has to create symbiotic relations with the natural system. The care of the earth is our noble obligation, which calls us now to preserve the land from destruction.[6]

We must live in and make use of the environment, at the same time preserving, as much as possible, environmental integrities. We can build the foundation of promising futures because we are endowed with a spirituality of responsibility for caring. However, Dubos warns us that humankind and the wilderness (forests, mountains, prairies, marshlands, deserts) can survive only if we create conditions, including laws and regulations, that retain the essence of both humanity and the wilderness. To maintain this essence is the noblesse oblige, the essence of caring for each other and for the earth, both now and for the future. Our obligation is to maintain the diversity of species—animals and insects. Our noble obligation is to make our inventions in nature regenerative and ultimately enhancing.

Dubos maintains that the earth's potential is unexpressed until used by humankind.[7] But love is required, not a spirit of conquest; nurture is needed, not domination or control of nature. We must respect the integrity, dignity and beauty of nature. Aldo Leopold said that "a thing is right when it contributes to the integrity, beauty and harmony of the biotic community. It is wrong when it goes the other way."[8] This is the basis of an agricultural ethic. Dubos asks the question, How much wilderness needs to be preserved in order to safeguard its integrity, future, and evolutionary process?

Dubos contends that humans can improve nature by using imagination, intelligence, and love.[9] This can happen when people have a larger purpose than profit and a strong sense of their obligation to the land. We ought to work with nature to enhance and preserve the environment over long periods of time. If the ongoing relationship between humanity and the earth is not considered, the design is not good enough. Dubos reminds us that "our knowledge about nature is only useful if it is always measured by long-range consequences."[10]

The suggestion is clear. Our science, technology, and industry must become ecologically based. In the modern world, science gives us the means to do many things. But it does not provide goals. Too seldom is it self-critical. In so many ways modern life echoes the lament of Captain Ahab in Melville's *Moby Dick*: "All my means are sane, but my purposes and goals are mad."

In today's world, with today's agriculture, we measure success by the latest statistic about the GNP, or by yields per acre, price per bushel or pound. But in the meantime, what is happening to the environment, *our* environment? Since ecology, the study of living things and their environment, is not ethically neutral, it presses the issue of quality of relationship, a fundamental value, which involves the health of and respect for the land. The quality of relationship between humanity and the earth, in the present and for the future, is inherent in the idea of noblesse oblige.

Dubos sees hope in the natural resilience of the earth, in the human ability to act to avoid crises, and in human ability to change. Humankind does not always remain as a passive witness to situations that are dangerous or unpleasant. Human intervention can alter the course of events, whether this involves the sheriff's office and town codes or decommissioning all nuclear reactors as the Swedish government is doing during the closing years of the 20th century.[11] Animals can, in many instances, adapt to change. But only humans have the capacity for conscious, planned change. We do not always change when we need to, yet the potential for creative change is always there. At this moment in our history, we must change from exploitation and destruction of our environment to the care and enhancement of that environment.

Dubos observes a high level of resiliency in the natural system. Indeed, we have observed that within limits, nature does have a tendency to bounce back. The squirrels and birds have returned to our hillside retreat. Over a period of many years, rivers and lakes can cleanse themselves if given a chance.

Implications for Agriculture

Agriculture has much to contribute to earth caring. Agriculture can be designed to maintain diverse species, which in turn maintains ecological stability that is mutually beneficial for humans and other living things. Tomorrow's agriculture can maintain the "essence of wilderness" and perhaps even enhance it. But, for this to happen, an ecological approach to agriculture (and its partners, industry and the economy) is needed. In other words, the health of the land is the concern of the entire society. The pursuit of wealth becomes a secondary matter. That may sound idealistic, but in truth it is a matter of survival. Economic stability of course is necessary, but not the accumulation of wealth at the expense of life itself.

The attitude of caring for the earth can be called agroecology (regenerative agriculture). Ethically, agroecology calls for our participation within the natural order rather than exploitation of that order by using water, soil, plants and animals, oil and gas to the point of exhaustion. Building such an agroecology is an essential social and ethical task. An agroecology demands an accurate definition of community, one that includes those things we really depend on. In an agroecology, human freedom is not the freedom to exploit but the freedom to nurture, to care for, and to enjoy the earth.

The resource base of our planet can perhaps support the present momentum of growth in productivity for another century. The toxicity in our environment has not yet exceeded the absorption capacity of the biosphere. We still have enough soil in most places to produce needed food and fiber for some time to come. Crop yields are still higher than past centuries could have imagined. Many fortunes can still be made from the land: but for how long and at what cost? These questions are not considered when the Gross National Product is calculated, although the GNP is thought by many to indicate our social welfare. If caring about the health and integrity of the earth and its future has no place in our science, technology, politics, and economics, we risk losing not only quality of life but life itself—our own life. In the classrooms of our colleges and universities, a new generation prepares not only to inherit the past but to shape the future. The possibility of regenerating and even enhancing the present resource base of our existence is exciting. Why not integrate the disciplines of landscape architecture, agricultural engineering, and the humanities (theology, language, history, fine arts, ethics, philosophy, political science) into the art of agroecology?

The Measure of All Things

Humanity is not the ultimate measure of all things. Perhaps the health of the earth is the measure. Not until very recently has the health of the land been considered a serious issue. Up until this point, the needs of commerce and military have prevailed. Now, however, with the advent of modern communications, we know that the health of the earth is threatened. We have a growing science in biology, human ecology, earth and atmospheric physics, and meteorology that previous generations did not have. But information alone cannot save the planet. As a Soviet official in this era of *glasnost* remarked, we will probably never communicate with advanced life on other planets, because even if such life existed, by the time it became sophisticated enough to communicate with other planets, it would have destroyed itself.

Technological advances make the question, Why should we care? even more pressing. The question calls not only for innovative technology, but for science at its best—self-critical science. New, fundamental commitments, not just from farmers but from all members of society, are needed. We need to see that the earth is not a commodity to be used up but a living and responsive thing, whose essence is beauty.

As one ponders these issues, one senses an even deeper question. What is the purpose of human life? This question is as old as humanity itself. While serving in Africa, I found this question of humanity's purpose and relation to the earth a topic of conversation almost everywhere. Yet in my own Western culture it is seldom voiced.

One cannot address the farm crisis, or for that matter the crisis of agriculture everywhere, without asking about human purpose. One possible answer, indeed the prevailing answer in our time is that we exist to make money and to own things. One of the former secretaries of agriculture, Earl Butz, understood agriculture as an industrial activity for generating wealth. As history has taught us, unless and until the whole social order supports an agriculture that preserves and enhances natural resources, the social order cannot endure. Is humanity the measure of all things? Or is the health of the earth and its future the real measure? Is humanity a partner with God in caring for this fragile planet or is the "wilderness" simply a resource to be consumed?

3

The Ecological
Frame of Reference

The Challenge to Change

History helps us see that agriculture determines the future of society. In our modern world, the way agriculture is done will very much determine the destiny of this planet. If annual monocropping prevails in relatively fragile areas, desertification will continue to spread; when the landscape is exhausted human society will come to an end. This is the reason for the title of this book: *Global Dust Bowl: Can We Stop the Destruction of the Land before It's Too Late?* Industrialized agriculture assumes that it must control and manipulate natural systems. But has this approach worked? Loss of topsoil, declining soil fertility, species extinction, high energy costs, loss of beauty, and the decline of the farming community all give reason to question the assumptions of contemporary agriculture. Agriculture should be designed to foster a cooperative, nurturing, and even enhancing relationship with the natural system. Instead of trying to control natural systems, we must learn to cooperate with those systems. How we view the world determines how we address the problem.

Our Worldview

A worldview is the scaffold upon which we construct reality. It is the loom upon which we weave our experiences in order to make sense of them. Our worldview shapes our relationships and experiences, including our science, technology, and industry.[1]

A worldview functions as a rationale for what we do. It shapes both our questions and our answers. Rather than change our worldview, we may try to find reasons why things are not working—reasons that will shift the blame to someone or something else. For example, we can blame the farm crisis on the balance of international trade. Our imagination conforms to our worldview.

Our present worldview was born during the Enlightenment, when humanity struggled to liberate itself from disease and famine. In earlier times we struggled to be independent of nature. To achieve this status, we had to try to conquer and control the world about us. The world was experienced as alien and threatening. We tried to separate ourselves from nature in order to transcend and control it. We separated the moral from the natural, and the modern age was born.

The modern worldview understands reality in terms of cause and effect. Nothing is unpredictable. Everything can be reduced to its component parts. Even the human imagination is reduced to the movement of the subatomic parts of molecules of the brain cells. Nothing exists but matter and energy. Nonmaterial things, such as norms and values, are simply names that conceal ambition. In this reductionistic and materialistic determinism, we can know only what comes from our physical senses. Since values and norms cannot be verified, they do not exist.

Modernism claims that the only route to knowledge is the physical sciences. Modernism reduces morality to the survival of the fittest species, individuals, corporations, or nations. Western modernism values radical individualism rather than the public good. Relationships are understood in terms of competition and conflict for survival rather than cooperation, nurture, and mutual support. Life, in the secular Western world, is competitive. Power, success, and wealth are prime values, but few people ask at whose expense power, success, and wealth are achieved. Meanwhile beauty, integrity, harmony, enhancement, and the good are reduced to sentimentality, which has no place in the "real world." Control and acquisition are the watchwords in our culture—we are defined as "consumers" and seem to placidly accept that dehumanizing, reductionist definition. We have lost our sense of reverence for the earth, and along with it our sense of dignity as human beings.

In agriculture, the survival of the fittest presently is the name of the game. A recent secretary of agriculture advised farmers to "get big or get out." This attitude has destroyed rural communities. Economics forces us to conform to the point that we are unable to evaluate even our agricultural science, technology, and industry. How can we break out of this destructive worldview? This question is being raised wherever we see our society on a course of self-destruction. The failure of modern agriculture to sustain the rural community and to care for the land is strong evidence of the inadequacy of our current worldview.

The farm crisis also raises doubts about the moral integrity of our culture. It suggests weaknesses in agricultural science, technology, and industry. It suggests a weakness in social understanding about the land and the community of those who cultivate the land. It indicates that there is social and economic weakness in the rural community and agricultural infrastructure. But this should come as no surprise since we design our science and technology in ways that are consistent with our worldview. Our culture understands the earth as an object to be used. We think of the land as capital to be spent for whatever purpose we invent. Our science, technology, and industry are organized to fulfill every desire, from genuine human needs to the most unreasonable and frivolous wants. Much of science, technology, and industry

37

is the result of a utilitarian, mechanistic, and reductionistic view of the world. Our prevailing agricultural science and technology are inadequate in view of the long history of environmental destruction and social injustice. New agricultural science and technology are needed. But scientific and technological change depend first upon a change in worldview and second upon a broadened understanding of the human purpose.[2]

An Ecological Frame of Reference

This book focuses on relationships. Human purpose is found not in control or acquisition but in relationships. Morality is life-oriented. Economics is for the common good rather than for private gain, though individuals may indeed benefit. Education is multidisciplinary and includes the study of values and norms (there really is no such thing as value-free education) and fosters awareness of interdependence. This book differs from other contributions to post-modernist thought, by having an ecological frame of reference. It points to new ways for humans to relate to nature and to one another as a whole, instead of the modern quest to dominate, manipulate, and control. Ecology, the study of the interrelationships of living things to one another and their surrounding environment, is central to this kind of postmodernism.

Of course modern, reductionist science would critique an ecological worldview as "unrealistic," and an ecologically oriented agriculture as "unmanageable." More than 20 years ago, my instructor in dissertation research said, "Keep your research manageable." Control was primary, relevance secondary. Not that the two should be mutually exclusive, but if we always try to keep things "manageable," we may never ask questions that need to be asked. Obviously, I ignored my adviser's counsel. But this is what postmodern thought must do: it must find ways to work with the complexity of an ecological frame of reference. It is better to face the complexity of new ways of thinking than to continue in old patterns that have proved to be destructive.

Our worldview influences what we value. A value is something that we desire, as the young boys on trail bikes valued power and speed. A value is an object of our greatest concern and consideration. We organize our lives and work around our values, which give us meaning and hope.

For example, by the time I was seven years old I had learned to value an automobile. My culture and family taught me this. Our family needed a car to get my father to and from work. In the neighborhood where I grew up, the make and model of one's car expressed one's social status. "Wheels" also meant freedom to do special things. I worked and saved, looking forward to the time when I could buy my first car. I wanted a new car, not a used one. More status! By the time I was seventeen, my earnings from weekends and summer vacations over all those years at last paid off. As a high school senior I drove a new Chevy and enjoyed the status. But five years later when I began my graduate studies, I sold it. The car got in the way of what I wanted to do. I no longer needed the status or the private sense of freedom that a car can bring. Cultural values shape personal values, yet personal values can change. Our spirituality and sense of purpose are reflected in the values we pursue. Our goals are attempts to actualize the things we value.

The dominant values in our culture are wealth and individual autonomy. In pursuit of these values, we develop a way of life that attempts to dominate and control. Competition and conflict, sometimes even to the point of violence, are socially acceptable. In the words of my neighbors down the road, "Why should I care?" or worse, "Boys will be boys." The purpose that drives our culture is owning things and controlling them. Economic growth is the measure of "progress" in modern Western culture. No wonder that our science, technology, and industry are made to serve these values. Likewise the purpose of corporate agriculture is the pursuit of power, wealth, and individual autonomy. But at whose expense? At what price?

In contrast, ecological values are quite different. Instead of power, wealth, and individual autonomy, there are the values that we think about from time to time but seldom take seriously: community, harmony, and beauty. Instead of competition, domination, and violence, cooperation and nurture could be valued. "Being" could replace "having." Theologian Alfred North Whitehead said that experiences of beauty can become goals of life and work.[3] Not mobility but stability is a measure of well-being in a postmodern ecology. Not consumption but regeneration and even enhancement are the goals. E. F. Schumacher wrote *Good Work*[4] and *Small Is Beautiful: Economics as if People Matter*.[5] He served as chairperson of the British coal board for 25 years, working in the area of energy production for modern industry. Schumacher was convinced that we must work toward a postpetroleum world, in which agriculture, industry, and transportation use only renewable resources. He wrote,"The most urgent need in our time is to ask the ancient question once again: What is the human purpose?"[6]

In the West today we lack consensus about our human purpose, both individual and corporate. During the Eisenhower administration, there was an effort to define the national purpose. After three years the project failed to reach a satisfactory conclusion. To what extent does our sense of purpose (or lack of it) shape our agriculture?

Recently there have been encouraging signs that care of the earth is becoming a common theme. On January 3, 1989, *Time* magazine's cover story was "The Planet of the Year: Endangered Earth." At the same time (December 1988) *National Geographic's* cover story was titled "Can Man Save This Fragile Earth?" In September 1989, *Scientific American* published its special edition titled "Managing the Planet Earth." Discontent about the direction in which we are moving, as described in these relatively widely read journals, is a first step toward change.

Patterns and Interrelationships

In contrast to modernism, which controls nature by reducing it to its smallest component parts and manipulating these parts, postmodernist ecology seeks to understand the interrelationships between living things and their environment. An ecological approach looks for patterns and associations and the roles living and nonliving things play in a system or habitat. Ecological thinking does not dwell too long on any one animal or plant apart from the others. Instead, ecology looks for how living things are connected. Humanity is within this range of consideration, not outside it. Human ecology studies the interrelationships within the whole of the land, or using Dubos's word, "the wilderness" which includes climate, soil, forest, streams, rivers, seas, and all living things and their patterns of sustenance. Within an ecological approach, everything is interrelated, like the strands in a spider's web. One thing is clear: every life (or species) is important. Every life contributes to the whole.

In nature, one species is not supposed to be more important than another. In fact, when one form of life dominates a system, the system becomes unstable and destructive. For example, some parts of Australia are teeming with rabbits that eat all the plants in sight. The rabbits, which are not native to Australia and have no natural predators there, have now multiplied into a plague of almost biblical proportions. Another example is the kudzu vine which, after being artificially introduced, is now growing uncontrollably in some parts of the southern United States. Yet a third illustration of this point comes from the Sahelian regions of Africa. Large numbers of cattle were allowed to overgraze the landscape when climatic conditions were favorable. Then during the drought period of 1968 to 1973, the situation became critical. The stressed landscape deteriorated to the point of being unable to sustain life.

So an ecological paradigm looks first of all for community, be it a plant, animal, or human community. It looks

41

for those things that nurture or contribute to the health and integrity of a complex web of interrelationships. Every organism has to establish its niche within the natural system, with the land. Otherwise it cannot survive. Within its niche, each living thing must not only survive and reproduce but contribute in some way to the welfare of the others. For an ecosystem to work, there must be equilibrium within the entire community. Postmodern thought, working with this ecological frame of reference, focuses on the interrelatedness of life and tries to respect and maintain life's equilibrium. The dynamics of this interrelatedness is more than matter and motion. Human will, which is expressed in actions, profoundly affects the world's many environments. Human imagination and hard work can restore and even enhance the natural system. Beauty cannot be quantified or marketed, but it is the product of a healthy natural system.

A timeless illustration is Chief Seattle's message to the great chief in Washington D.C. when tribal land was sold in the mid-1880s. In it we glimpse other ways to understand our relation to the earth. "The earth does not belong to people; people belong to the earth. This earth is precious to the Creator and to harm the earth is to heap contempt upon its Creator.... Our dead never forget this beautiful earth, for it is the mother of the red people. We are part of the earth, and it is a part of us."[7] In his letter, Seattle wrote that "every part is sacred" and that "we are all connected." He asked, "What are men if all the beasts die?" He knew that "whatever happens, happens to all, for we all share the same breath." He said, "Tell your children. Teach your children that the earth is our mother and whatever befalls the earth befalls the sons of the earth." Then he asked why the white people wanted to buy the land and how such a thing as land could be sold. Although no answer was offered, he left us with a disturbing challenge that has yet to be addressed: "If we sell the land to you, love it and care for it with all your strength and heart. Preserve it as God loves us."[8]

How then shall our agriculture shape its ecological niche within the "wilderness" (or upon the land), so that it maintains an ecological equilibrium with other forms of life? This question points to new agendas for research and development that go well beyond the "production efficiencies" that we know are counterproductive and, in the long run, inefficient. We must find ways to regenerate the land upon whose life our lives depend.

It was in the nation of Sarawak, on the island of Borneo, that I began to think about the need for a new "blueprint" for agricultural development. During the heady days of the Green Revolution I was involved in programs to expand the production of the new high-yielding rice then being developed at the International Rice Research Institute at Los Baños in the Philippines. These varieties were developed to be highly responsive to modest inputs of nitrogen fertilizer. The idea was to increase yields.

During the course of my work I began to observe that farmers had abandoned the older time-tested varieties of rice, which had lower yields but also had acceptable resistance to insects and diseases. Also, to take advantage of the new technologies, more land was devoted to these rice crops. Fruit, nut, and pepper plantings were neglected. At the peak seasons of planting and harvesting, family routine was disrupted. The old peasant farming systems were breaking down. So were human relationships. Chemical pesticides had to be used. The residues caused many problems that were unforeseen in the beginning. In programs to extend the use of these new "miracle" rice varieties, people, their communities, and other farm activities began to fall apart. Today many of these problems have been resolved. But in the early days of these innovations, I began to learn of the necessity to think about relationships and what happens when they are shaken.

The Dynamics of Social and Scientific Change

Thomas S. Kuhn in his classic study *The Structure of Scientific Revolutions* developed a very helpful profile of the way scientific change occurs.[9] Kuhn observed three fundamental elements in scientific revolution, whether it involved mathematics, astronomy, physics, or agriculture. First, a crisis necessitates the community's rejection of a time-honored theory in favor of another that is more compatible. (For example, this book, *Global Dust Bowl,* argues that agroecology is more compatible with our needs and purpose than the present state of agriculture.) Second, each crisis or scientific challenge points to new areas for scientific scrutiny and new standards of evaluation. Third, such changes, together with the controversies that almost always accompany them (this book contributing to the controversy) are the substance of scientific revolutions.[10]

In the case of agriculture, the seeds for change lie in the debate about what compromises or constitutes efficiency. Because the educational process in agriculture and other sciences is both rigorous and rigid, the answer to the question of efficiency often reflects the prevailing worldview that exerts a strong influence on the scientific mind. Thus research becomes a strenuous and determined attempt to force nature into conceptional boxes designed by professional education. We may even wonder whether research can proceed without such boxes, whatever the element of arbitrariness in their development. Yet one must also ask whether the scientific community really knows what the world is like and what it needs. Too often success within the scientific community is based on its ability to defend its assumptions and to suppress any novelty that threatens basic commitments to theory and development. I urge the reader to consult Busch and Lacy's work, *The Politics of Agricultural Research.*[11] This book gives breadth of understanding within the agricultural sciences in regard to Kuhn's analysis. That in turn enables us to understand the dimensions of the problem of Western industrialized agriculture.

Our science searches for conformity. Science "fails" if this conformity does not happen. And the scientist fails rather than the theory. As Kuhn says, "A problem is to be solved within the paradigm reference."[12] In the case of agriculture, we return to the prevailing assumption that there are problems to be solved within our agriculture rather than asking, Is our agriculture the problem?

Agriculture has social and political implications as well as scientific ones. For example, some people think that environmental concerns distract us from the greater problem of social injustice. Until the recent farm crisis, few people saw the connection between environmental and social issues. The modern penchant for compartmentalization and specialization has blinded us to larger interpretations. We can't see the forest for the trees.

Kuhn points to the role that crisis plays in social and scientific change. If there are enough "surprises" or anomalies, we gradually recognize that the prevailing views are inadequate. "A profound awareness of a problem is basic for paradigm change."[13] Indeed, one can observe some professional insecurity among scientists, farmers, and agriculturalists and a growing recognition that the existing rules need to be changed. Kuhn points to a further element of change, namely that when a crisis is not resolved, alternative solutions begin to get a better hearing. About this he says that we are unlikely to change our way of thinking until a crisis demands it. A crisis involves the realization of danger and it demands more serious attention, but it also presents an opportunity for change. Throughout history when pressured by crisis and social demand, science has searched for promising alternatives.

The history of scientific revolution teaches that more often than not, fundamental inventions are made by people who are young and relatively new to the field. Invention comes from people with little commitment to the old ways and traditional rules, and who, in the process, design new rules for the game.[14] In some instances the result has been a revolution in science!

Confronted with anomaly or crisis, scientists take a different attitude toward existing paradigms, and the nature of their research changes accordingly. The proliferation of competing articulations, the willingness to try anything, the expression of explicit discontent, the recourse to philosophy and to debate over fundamentals, all these are symptoms of a transition from normal to extraordinary research.[15]

The design of agriculture has been and will continue to be an effort of the entire social order. Its design reflects the values of a given culture. It reflects the dominant worldview that has evolved within the culture. What happens to the land is directly related to the social rules and norms the culture has designed. But what happens to the land also shapes the social ethos. An eroding landscape erodes the social ethos. It raises the question of the purpose of the land and society and of the objectives of the science and technology of the society. Caring for the earth involves the ecology of land, agriculture, and society—the interrelatedness of all life. None of these things can be understood apart from the others.

It is indeed possible to care for the earth. It is encouraging to recognize that there are nearly 20 land-grant colleges searching for new faculty to work in the field of regenerative agriculture. And there exist many significant private centers of activity where research and training is being done.[16]

Added to these many institutional innovations are more than 40,000 registered organic farmers and the emergence of new marketing systems to accommodate their produce. The farm crisis is necessitating the community's rejection of a time-honored theory in favor of another that is more compatible with the land and the people on the land. The scientific community in a few significant places is looking at new problems with new sets of criteria and standards, and there is indeed great controversy about whether agriculture has problems to be solved or whether agriculture itself is the problem. The outcome of this debate has everything to do with how the question in the subtitle of this book will be answered.

4

What the Forest Teaches

Mutually Enhancing Relationships

Three years after my wife and I began our work in agricultural development assistance in the Belgian Congo, in 1956, I observed that the corn yields in one field had fallen considerably. Late one afternoon, a local villager saw me walking the rows, trying to discern some reason for the decline. The field had no weeds, insects or disease. I had used a reliable local seed. Certainly our land preparation was adequate; I had applied about 150 pounds of nitrogen per acre. I had used the same methods that had worked in previous years, both on the plot where I was standing and for adjacent fields. But the yield was marginal. The villager took one look at the field and then looked at me. "Things do not look very good," he said. I agreed and then asked him what needed to be done. He replied: "Go into the forest and see what it teaches," he replied. Then he left me and continued on his way. I did not understand his advice. Six months later I was still pondering his remark. I began to spend some time in the nearby forest, the high bush country of the Katanga to see if I could fathom the villager's remark. It took five more years before I felt that I truly understood.

In the tropical forest, the soil is sheltered from the sun, heat, and tropical downpours of the rainy season. The soil temperature remains relatively constant. There is no erosion. Organic matter accumulates and decomposes at a steady rate. As I scratched around with a small African hand hoe, I discovered that all the trees and shrubs had extensive surface "feeder roots," all within the upper half-inch of the soil. These feeder roots are very small and fragile, like spider webs, for catching soluble nitrogen when it becomes available. Below the feeder roots, which penetrate the decaying forest leaves, is a large support system. Some of these roots go down nearly 40 feet, reaching for moisture during the six- to seven month-long dry season. Beneath the decaying vegetation at the soil surface, everything was moist and cool. This layer teemed with insects and fungi of every description. Above me, I could hear birds singing, occasionally joined by sounds of rodents and reptiles. The whole forest was alive. It was a very different place than the field of corn where only one thing grew and everything seemed—or had seemed—so predictable.

In the cornfield, the soil was red. In the forest the upper half-inch of soil was dark brown to black due to the concentration of organic matter. But the soil in the cornfield was beginning to crust. This was my first experience of "laterization," the crusting, sometimes to stone-hardness, of geologically ancient soils with rich concentrations of iron and aluminum oxide. Laterization does not happen if there is abundant organic matter in the soil. But if the forest (or grassland) is cleared for annual cropping, the whole system soon collapses. The process can result in desertification.[1]

The forest taught me that soils are very fragile. They can be cleared and planted in grain crops only at great risk and for short periods of time before soil is damaged and lost. The forest teaches that soil cannot be exposed to full sunlight and driving rainfall (or in northern and southern latitudes to rapid snowmelt) without harm. The forest teaches that plant, animal, and microbiological life all depend upon the organic material which is the basis of life. The forest teaches that relationships are complex and cannot be simplified without damage. The cornfield destroyed the integrity of the natural systems in the region where I lived. After only three years the cornfield had failed to produce a significant yield even though nitrogen fertilizer had been applied. The advice of our neighbor (who never learned to read or write) was the best advice that I had ever received in my life. "Go into the forest and see what it teaches." In agricultural colleges in the 1940s and 1950s, I had studied the chemistry and physics of particular plants, insects, reptiles, and rodents. I had never thought about how these interrelate and neither, perhaps, had my instructors.

The forest teaches us that to speak of land is to speak of life in its fullness and complexity. Land is more than soil. It is more than acreage. It is more than real estate, territory, or raw material. Indeed, we belong to the land and share with it a marvelous, mutually enhancing relationship. In the peacefulness of the forest, we sense once more the awe and majesty of our belonging to the land.

Our son and daughter-in-law live in Senegal, West Africa, and work on issues related to the environmental crisis of the Sahel. While exploring the expanses of dry land, our son came upon tree stumps in the desert, dramatic reminders that this region had once supported life well. In the 19th century, there had been dense concentrations of trees near the riverbeds and plentiful grasses in the savannas. Many animals and birds lived there, including giraffes, ostriches, antelopes, and elephants. Through the years trees were cut for gum arabic and to clear the land for cultivation. Then wells were dug, commercial livestock interests moved in, and the traditional patterns of the native animals, birds, vegetation, and water were disrupted. The drought years of 1968 to 1972 severely compounded the crisis. Most of the remaining trees died or were cut for fuel or wood for construction. Without adequate vegetation, many animals and humans died. The Sahel was becoming a desert. All around was evidence of devastation that had been greatly increased by humanity's misuse of natural resources.

Modern Western society assumes that the land is resilient and abundant. We think that the land is tough and that it rebounds after stress. And in the United States, we still think that the land is a frontier to be conquered. We learned this as our parents read bedtime stories of the adventures of Paul Bunyan with his giant ax and plow. We see movies about new lands to be conquered; even *Star Wars, Star Trek,* and other "space movies" are extensions of this theme. Although the frontier is gone, the old assumptions prevail in personal attitudes and social policy, including farm policy. Remaining forests continue to be logged without adequate care or replanting. Fragile prairie and desert lands continue to be plowed and irrigated. Prime agricultural land continues to be converted permanently to nonfarm purposes (how many more shopping malls do we need?), and desertification continues unabated. It is long past time to ask, Is the land fragile? Are there limits to its resilience? Our society must ask how we can protect the land and perhaps even enhance it. Can we stop its destruction before it is too late?

These are foundational questions. Yet despite the farm crisis in the United States and nutritional and ecological problems all over the world, the questions are seldom raised. The farm crisis is a crisis of culture, its sciences, technology, and industry, and of our self-understanding as human beings.

The Planetary Miracle: Community of Life

The planet earth is relatively small. By our standards it is extremely large, yet the biosphere (that slice of the earth and its atmosphere which can support life) is so thin that it is virtually transparent. If we could reduce the earth to the size of a billiard ball, its mountainous surface would appear ivory smooth. We might be able to recognize the continents, but the oceans would be mere films of dampness and we would not be able to detect the biosphere at all.[2] The astronauts who looked back at our beautiful planet (our only known haven in what appears to be endless, inhospitable space) saw only the brilliant blue and white of water in all its forms. This is our most precious resource. Ninety-seven percent of this water is in the oceans. Of the earth's fresh water, more than 99 percent is retained in the ice caps and glaciers or is underground. Less than 1 percent is in lakes, rivers, and soil and is available to support human life and its technologies for producing food and fiber.[3]

Had the planet been slightly closer to the sun, its water would have remained in the form of vapor. Had it been slightly farther away from the sun's heat, its water would have remained frozen. No circumstance has greater significance than earth's critical distance from the sun, which allows for the thin range of 100 degrees centigrade within which water can exist as liquid.[4]

Biologists have long pointed out that the life that exists on earth could not exist on other planets where intense heat and freezing cold prevail. From an astronaut's point of view, the temperature range of 100 degrees centigrade (30 to 120 degrees Fahrenheit) in which active life is manifest is incredibly narrow. The adaptation of various organisms to this limited temperature range is further effected by the alteration of day and night and of the seasons; there are also elevation (sea level to mountain top) and latitude to consider.[5]

Thus from a few miles up in the atmosphere to several feet under the earth's crust and into the depths of the ocean lies what we call the biosphere, where living things are engaged in the continuing drama of reproduction, consumption, decomposition, and renewal. About three-quarters of the breathable atmosphere consists of nitrogen, while the rest is mostly oxygen. The extraordinary thing about that part of the atmosphere which can support plants and animals is its thinness. Earth is about 8,000 miles in diameter and 25,000 miles in circumference, but there are only about seven and one-half miles of gaseous exchange above its surface. When puffing at the summit of Mount Everest (29,141 feet), or having a meal in a jetliner at about that same altitude, the air pressure is below one-third of that at sea level. One cannot breathe without artificially supplied auxiliary oxygen. The drylands or continents represent but .4 percent of the earth's total mass. It is upon this surface that plants, animals, and humanity live.[6] Slight and delicate though it is, the biosphere is sufficient for every life purpose. It is miraculous and unique in the cosmos. But it is also very small, this land upon which everything we know exists.

Within this very thin and nearly invisible biosphere, each living thing depends upon the constant flow of energy from the sun. This energy must pass through three fundamental stages or trophic levels. One consists of the green plants which receive energy from the sun and fix it in the form of foodstuff for themselves and for all other living things. This level includes, of course, aquatic phytoplankton (microscopic marine plants). The second level consists of all animal life. No animal can fix solar energy. Animals must eat either plants or other animals that have themselves eaten plants. Thus plants are producers of energy, and animals are consumers of energy. The third trophic level consists of organisms called reducers, or the microscopic life in the soil such as the fungi and bacteria which break down the structures of dead plants and animals and which release their components to the cycles of the whole biosphere.[7] This interdependent cycle of life has been in existence since long before humanity came on the scene. Indeed, one might say that it could have all gone on quite well without us.

This cycle is not fixed and unchanging, but growing and developing. Evolution is the perpetual exploration of new environmental opportunities. Life as a whole may be described as opportunistic. For example, one might picture a lichen, or perhaps some small vine, as representative of life. With its tiny shoots or tendrils, it is forever exploring new cracks, new crevices and niches in which it may fasten the tip of a searching growth bud and there perpetuate itself.

There are in the world today about 8,600 distinct species of birds. There are about 5,000 mammals, about 6,000 reptiles, some 1,500 amphibians, and more than 20,000 fishes. Thus the existing number of vertebrate species is something over 40,000, of which humankind is one. Numbers of invertebrate species are astronomical. There may be more than a million different kinds of insects, to say nothing of their innumerable relatives. We humans represent only one form of life among millions.[8] We are not the only earthlings. The diversity of life on earth is the key to the stability of the land. This insight must not be forgotten.[9]

Perhaps a good illustration of all this can be seen in the life of the simplest community: the lichen. There are many kinds, most of them growing in such bleak, inhospitable surroundings that they may be the only sign of life. Lichens manage to eke out a living in some of the most biologically impoverished parts of the world because they are not a plant but a community. A community is more than the sum of its parts. Lichens are formed by a symbiotic relationship between two different kinds of plants: fungi and algae. The fungi contribute something and the algae contribute something and together they both thrive in what may be the world's most basic community. Lichens come in all colors and shapes, but each is a partnership in the very best sense. The fungus provides the foundation by secreting acids that help the lichen to maintain a foothold and draw mineral nutrients from the rocks beneath. The algae, in turn, feed the community. Algae have chlorophyll, the vital substance for the process of photosynthesis. Together, the fungus and the algae endure where neither could exist without the other.[10] The lichen works remarkably slowly, but gradually, over thousands of years, it breaks down the rock base into soil fragments, thus making the rest of plant and animal life possible.

One cannot have a sound perspective on human history if the role of soil is ignored. Nor can one properly interpret today's domestic and international problems without taking soil into consideration. We should say "soils" in the plural, for soil exists in great variety, and this variety has been significant for plants, animals, and the human species. The relationships are intricate. Without living organisms such as bacteria, fungi, and lichens, there would be no soil and no life as we know it. The stuff that covers the earth so very thinly becomes soil only by being lived in and upon! It comes about by the interaction of air, water, light, and life upon the rocks. The energy for this complex operation comes from the sun.[11]

Just as coal, gas, and oil are the legacy of past sunshine stored below the surface of the earth, so is soil. Soil, the dark carpet upon the earth, is far more precious. The generations of plants and animals that form the soil are sustained by it during their brief existence. What remains, as evidence of their activity in catching and transforming the sun's energy, is the organized soil. Vulnerable to misuse yet vital for our existence, the soil is generous and responds to good cultivation.[12]

Consider for example the deepest and most fertile soils on the earth's surface—the prairies of the North American continent. Thousands of years of growth and decay of the heavy sod of indigenous grasses resulted in deep, rich soil. Such soil does not develop under the trees of the forest. When the trees are removed from a forest there remain but a few inches of topsoil at most. But on the prairie, from one to two feet of accumulation of fibrous roots, estimated to produce more than four tons of roots per acre (some reaching to depths of 12 feet), have resulted in one of the earth's most precious resources.

During the same time that the first forms of life began the process of soil making, the ecosystems that now occupy the oceans developed. They have existed longer than any other communities of life on the planet earth and they now contain an incalculable number and variety of delicately balanced living things. In addition to its plants and animals, the sea holds the minerals and other materials delivered to it from the land masses in quantities about which we can only guess. Its energy flows from phytoplankton (microscopic marine plants) to zooplankton (microscopic marine animals), through a bewildering array of lesser and greater invertebrates, fishes and mammals, through which the whole cycle of life passes back to bottom-feeding decomposers on its return to the chain of life.[13] From the ocean emerges most of the oxygen of the atmosphere; 99 percent of it from the upper 60 feet of the ocean's surface where the sun is the most effective in energizing the photosynthesis of the phytoplankton. From these same bodies of water and from lakes, streams, and ponds as well as from plant leaves and people, evaporation lifts about 100,000 cubic miles of water from the earth's surface every year for recirculation. Such is the force of solar power, and the interaction and interdependencies of the sun, sea, soil, animals, and plants.

Although the water cycle is by far the most impressive of the earth's life-support systems, there are others that are also vital to the maintenance of living organisms. These are the carbon and nitrogen cycles which make their unique contributions to the soil. Good soil is filled with multitudes of microorganisms, among them, bacteria and fungi. These organisms do not possess chlorophyll, so they must operate on chemical energy instead of solar energy. They are the power sources for the carbon and nitrogen cycles.[14]

The forests, especially in the tropics, offer the next widest variety of living beings in any given area. The typical equatorial forest consists of layers, from the great green canopy of the largest trees through the shade-tolerant forms that stand beneath them to the lesser trees, shrubs, and smaller plants of the cool and luxuriant floor. Each stratum has its characteristic birds, insects, reptiles, and mammals.[15]

Like all life forms, forests are always changing. Even without human intervention, a forest is constantly in a state of transformation. Trees grow to their fullest extent, mature, and then die. When at last they fall, the sunlight, streaming through the gaps left in the canopy, promotes the immediate growth of other species which until that moment had been inhibited by shade. Later, fallen logs nurture yet other species. New kinds of trees, new groups or associations of species are constantly striving for succession.

The forest and grasslands produce much of the earth's oxygen and maintain the balance of the earth's carbon dioxide. An overmature forest consumes as much oxygen as it creates, but a vigorously growing region of young trees each year consumes five to six tons of carbon dioxide and gives off four tons of oxygen per acre at the same time that it produces about four tons of new wood.[16] A young, healthy tree can have a net cooling effect, through transpiration, equal to ten room-sized air conditioners operating 20 hours a day. Under the canopy of a large tree, the air may be 20 degrees cooler than elsewhere.[17] The forest, dependent upon the soil, which is dependent upon the life and energy cycles within the soil, is one of the prime regulators of atmospheric balance and climatic stability. Survival is based upon the production of essentials for all of life. As one meteorologist put it, "The cause and effect links of the climate system, every place on earth, is connected to some extent to the climatic system in every other place. That is, a kick in one spot will cause a bulge elsewhere."[18] Indeed, agriculture has been a big "kick" at least since the Roman Empire; perhaps longer than that. The force of this kick increases in our time. And yet this kick, so very damaging, has happened during a brief moment of time in terms of planetary history. This history adds a much-needed perspective to the subject of earth caring and the ecology of land, agriculture, and society.

Imagine, for example, that the 4.5 billion-year period involved in the development of the earth could be described in a time line of one week. On early Sunday morning interstellar debris coalesced to form the earth; the rest of the solar system and the planet heated up, and outgassing from volcanos and rocks began (4.5 billion years ago). By Sunday evening about 7:00 P.M. (4.0 billion years ago) the earth's primitive atmosphere and ocean had formed. Lightning, sunlight, hot lava, and thermal springs sparked the creation of molecules in a slightly acidic ocean.

At 2:00 A.M. Monday (3.0 billion years ago) the earliest forms of life emerged, as we have traced them in the striped iron bands in the rocks of Greenland.[19] At 10:00 P.M. on the same day (2.9 billion years ago) early forms of bacterial colonies emerged as we detect them in the recent discoveries of fossilized remnants revealed in Australia. Tuesday, 8:00 A.M. to 4:00 P.M. (2.8 billion years ago), filamentous photosynthesizing bacteria flourished. On Wednesday at 9:00 P.M. (2.0 billion years ago) a remarkable explosion of diverse bacterial life took place, and in Canada today we have the world's first known evidence of Precambrian microfossils.

On Thursday at 8:00 P.M. (1.4 billion years ago) the first cells with chromosome-bearing nuclei appeared from which every plant and animal evolved, except for bacteria and blue-green algae. Friday at 10:00 P.M. (700 million years ago) the earliest invertebrates, resembling jellyfish, sponges, sea pens, and worms, were formed. Now the stage was set. At 1:30 A.M. on Saturday (600 million years ago) marine life began to flourish, with shell-bearing animals such as trilobites and brachiopods. At 9:00 A.M. on this same day (400 million years ago) jawed fish and small land plants appeared. Just before noon on this day (350 million years ago) amphibians marked the transition from water to land and insects later emerged onto a landscape dominated by impenetrable forests and swampland.

Late Saturday afternoon, at 5:00 P.M. (190 million years ago) early dinosaurs appeared, as reptiles continued their great expansion. Shrewlike mammals evolved at this time, while primitive birds armed with teeth arose 45 million years later. Then, at 10:35 P.M. on Saturday (1.6 million years ago) the oldest common ancestor (the Aegytopithecus) of the apes and human beings made its home in the dense forest. At three minutes before midnight (1.5 million years ago), Homo erectus walked the earth. Then, at 11:59:56 P.M. (four one-hundredths of a second before midnight, 30,000 years ago) Homo sapiens pursued art and religion.[20] During this time (10,000 years ago), we find the beginnings of agriculture and its effects upon the planet. In these final fractions of time, the post-Reformation, post-Renaissance, modern scientific, technological human beings appeared, and in addition to their many achievements, soil and forest resources have disappeared, deserts have enlarged, and the earth is covered with a blanket of dust and smog that significantly affects the temperature of the biosphere and the level of the sea. At this moment (for a moment is all we have), we ponder the question of earth caring and the ecology of land, agriculture, and society.

Promising Futures: The Call to Trusteeship

The perspective of evolutionary time has long been part of Western consciousness. This perspective is especially important to agriculture. Short-term cycles of investment and return result in eventual loss; it is destructive to limit our understanding of agriculture to a mere means for generating wealth. It is much more practical to do agriculture within the long-term context of geological and evolutionary time. Furthermore, to work the land with any other attitude but reverence and humility is destructive. On this point we need to recall the words of Liberty Hyde Bailey, professor of horticulture at Cornell University from 1888 to 1903, and then, until 1913, the dean of the New York State College of Agriculture and director of the New York Agricultural Extension Service. Reflecting about agriculture and agricultural education, he said,"Into this secular and more or less technical education we are now to introduce the elements of moral obligation. . . . This cannot happen until the farmer and every one of us recognize the holiness of the earth."[21]

Prior to this he wrote that farmers represent society in caring for the earth. Farmers are also the agents of the God who made us. Therefore farmers must exercise dominion with due regard to all these obligations. As trustees, farmers must work to increase the productiveness of the earth from generation to generation.[22] It seems to me that the word "trusteeship" expresses what our relationship to the land should be. This idea has a long history.

In the opening chapters of Genesis, that magnificent prelude to all the ancient texts, the writers say that humanity was created in the image of the Creator and given dominion over the land. Humanity was to fill the earth, and to be fruitful and multiply.[23]

In ancient times as now, the ruler or governor could only be in one place at a time. It was the custom to place symbols of the ruler in highly visible places to remind the people of the authority of those who reign over them (just as we place photographs of our presidents and governors in federal and state buildings). To say that we are created in the image of God means that the people are symbols and representatives of the Creator, with the capacity to create, imagine, remember, have compassion and vision, and to maintain order within a domain. In ancient Hebrew thought, justice or righteousness demanded right relationships of all life: human and nonhuman. In ancient Hebrew thought, justice was wholistic or comprehensive. To be "in the image" of the Creator was to be responsible for maintaining justice within God's domain. In ancient Hebrew thought, the ruler was accountable to God for the maintenance of this justice. The Jubilee Year was a part of this function. In Genesis 2:15 the mandate is to "till" (or cultivate) the land and to "keep" it. The Hebrew word for cultivate also means to serve. To cultivate the land is to serve the land. To keep the land is to care for it as a sacred trust. In the Old Testament, private ownership of land developed over a long period of time, but the concept of trusteeship prevailed at least as an ideal. But whether land was held privately or communally, it sustained human life. In many of the Psalms, Deuteronomic codes, and writings of the literary prophets, human welfare was tied to the welfare of plants and animals.

In the context of Hebrew thought about aristocracy, the ruler who failed to maintain justice (in the wholistic sense) forfeited the right to rule.[24] Fruitfulness was another important idea. To be fruitful was to contribute to the welfare of the community and the productivity of the land. The later concept of shalom, which appears in today's language, encompasses these wholistic notions. In justice and righteousness is peace. Peace is ecological.

The idea of trusteeship in our contemporary world is quite close to these ancient meanings of having dominion, of keeping, and of being fruitful. Today if a person is invited to be on a board of trustees, that person does not lay claim to ownership of, say, a college or hospital. Rather, a trustee accepts responsibility for the stability, welfare, and enhancement of the institution. When we ask how we ought to relate to the land, the idea of trusteeship, together with the Hebrew traditions of shalom and fruitfulness, provide direction.

Humility, Enhancement, and Trusteeship: Foundations of a Land Ethic

The feminist movement has reminded us of the power of language. Our choice of words reveals our attitudes about many things. The more I work in agriculture and ecology, the more I realize what lies behind certain words and phrases commonly used in agriculture. For example, I can no longer employ the term "land use" because I am in relationship with the land. It is not a thing to be used. We do not talk about "using" a friendship, a spouse, or a child. There we think in terms of relationship, not exploitation.

Humility is a fundamental attitude in approaching agriculture and agricultural education. If we evaluate what we do from the perspectives of biology and of planetary and human history, we can work with the land with an attitude of humility. God has entrusted us with the land. We are to be fruitful in that relationship, knowing that dominion includes not only power but responsibility. We discover new (or perhaps ancient) relationships that transform our self-understanding, renewing our sense of purpose and of life's sacredness.

Persons like René Dubos,[25] Liberty Hyde Bailey[26] and Aldo Leopold,[27] to identify only a few prophets in our time, suggest that good relationships are defined in terms of enhancement. Leopold says that a good relationship contributes to the integrity, beauty, and harmony of the biotic community. Bailey suggests that good relationships with the land result in a steady increase in land fertility from one generation to the next. Dubos speaks of quality and enhancement as measures of good relationships. All this implies that our prevailing concepts of land stewardship and land (soil) conservation are not good enough. We can do more than conserve available resources. We must do more than simply make a resource last as long as possible. An ecological ethic will seek not only to preserve the land but to enhance it. This carries us far beyond the idea of conservation, important as that is. We have come far enough in the development of the biological and earth sciences to realize that it is indeed quite possible to meet these high standards. This is particularly true in agriculture. Agriculture can be regenerative and enhancing. This ethic can move us far beyond the simple issue of production, opening new possibilities for practicing dominion in the biblical sense of that word. By now the reader should be well aware that the ancient meaning of dominion does not include domination.

In addition to humility and enhancement, the third element of a postmodern land ethic involves trusteeship of the land. Trusteeship needs to replace the idea of ownership. To accept trusteeship of the land is to accept a high honor offered by the whole society, an honor as great as being invited by a university or foundation to serve as a trustee.

The society structures itself to make the sharing and implementation of this responsibility possible. The trustee "keeps" the land in trust. The society respects and safeguards this role and responsibility in every way possible, so that land trusteeship involves a mutually enhancing relationship between the trustee, the land, and the society.

In summary, the foundations of a postmodern land ethic are: an attitude of humility about our relationship with the land, enhancement (which includes but moves beyond conservation), and trusteeship. When this ethic is working properly, it results in a relationship that enhances all species over a long period of time.

High Expectations for the Land's Future

There are several good reasons to hope that a new agriculture will emerge. First, there is a growing worldwide recognition of the environmental stress on our planet and the possibility of a global dust bowl. Every day the news media reminds us of hunger and famine that result from rural injustice, war, deforestation and desertification. Talk of global summitry to discuss environmental problems gains momentum.

The United Nations World Commission on Environment and Development in its report *Our Common Future,* issued this urgent call for achieving sustainable development by the year 2000. The report urges nations to (1) find ways to translate concern for the environment into greater cooperation among countries at various stages of economic and social development so that they may achieve common objectives that enhance relationships among people, resources, environment, and development; (2) consider ways the international community can deal more effectively with environmental concerns; (3) help define shared perceptions of long-term environmental issues and the appropriate efforts needed to deal successfully with the problems of protecting and enhancing the environment; (4) develop a long-term agenda for action during the coming decades; and (5) set goals for the world community.[28]

The chairperson of this commission, Gro Harlem Brundtland of Norway, said, in reference to the call to action outlined above, "We call for a common endeavor and for new norms of behavior at all levels and in the interests of all. The changes [must come] in attitudes, in social values, and in education, debate and public participation."[29] In March 1989 the first international environmental treaty was signed, banning the use of chemicals that deplete the ozone layer. Some countries did not participate, but this was an important step nonetheless, and an example of the type of change called for by the U.N. Commission. We are increasingly aware of toxicity in the food we eat, the water we drink, and the air we breathe. The problems of soil loss, freshwater scarcity, and species extinction are now public knowledge. The spectacle of farm bankruptcy indicates that society must find ways to stabilize the farming communities, and do so with integrity. We know that we are in an environmental crisis unprecedented in history. But in this crisis there is hope. The Chinese word for crisis has double meaning that includes both challenge and opportunity. For the Chinese, a crisis can result in a catastrophe or a breakthrough. When we realize that our survival is at stake, perhaps we can achieve such a breakthrough.

In the Worldwatch Institute's publication *State of the World, 1988,* Lester Brown wrote: "In preparing this annual assessment during each of the last five years, we have in effect given the earth a physical examination, checking its vital signs. The readings are not reassuring. The earth's forests are shrinking, its deserts expanding, and its soils eroding all at record rates."[30] He went on to say that "assessing these threats to the future can easily lead to apathy or despair." But later in the discussion, in reference to the international meetings held in Montreal dealing with chlorofluorocarbon emissions (which threaten the earth's protective ozone layer), he said: "These developments suggest a new era, one in which attention shifts away from East-West ideological conflicts and toward the reestablishment of an earth with stable, healthy vital signs. The world has come a long way from the mid-seventies when environmental concerns were considered something that only the rich could afford to worry about. Today, they are concerns no one can afford to ignore."[31] Indeed, it is reasonable to think that humanity is capable of changing its collision course with environmental calamity.

The second reason for hope is related to the first. It is the recognition that agriculture, with its petro-chemical and capital-intensive technologies, cannot be sustained if we wish to address the larger problems just described. Agriculture that achieves production "efficiencies" at the expense of the natural environment and the rural community is part of the crisis. Today's agriculture is not an analogue of the biotic communities in which it finds itself. Rather, it is an oversimplification, like that cornfield next to the forest in what was then called the Belgian Congo. Several nations have taken new steps for the welfare and future of their agriculture. Examples are discussed in chapter 6. There are several global institutes now in place that reflect the growing commitment to the land's strength and health, in the true spirit of trusteeship.

A third reason for hope about the future of agriculture is the growing recognition that the earth is a very precious and fragile planet. Evolutionary biologists, meteorologists, astronomers, astronauts, and earth physicists have contributed greatly to public knowledge about these matters. We now realize that even though "mother earth" cleaned up after us in the past, we are now on our own and must clean up after ourselves. Environmental science and the developments in biology and biotechnology hold great potential for developing agriculture that can rehabilitate stressed lands, preserving and enhancing them.

Finally, much work is already being done with alternative agriculture, especially in research for regenerative technologies and appropriate socioeconomic support structures.

Chapter 6 points to the frame of reference of these activities: an agroecology. Even the 1985–89 U.S. farm bill has some provisions (though far from adequate) for encouraging these new efforts.

"We know we belong to the land and the land we belong to is grand," wrote Rogers and Hammerstein in that great chorus for the musical *Oklahoma*. The tune and lyrics have been in my mind since the play was written in 1947. The more I work in the ecology of agriculture, the more I identify with this verse. Indeed we do belong to the land. The land does not belong to us. David Attenborough put it this way in his film series *The First Eden:* "Nature has taken care of humanity. The problem is, humanity has not taken care of nature." But perhaps the present crisis will impel us to do just that: to take care of nature, to take care of the land.

5

A Brief History of Agriculture

A Personal Encounter with History

During the Second World War, my father was superintendent for a chemical fertilizer factory in southern California. Because of the war, labor was in short supply. At age 15 I worked on weekends and school vacations as a stevadore, unloading bulk chemicals from freight cars and loading the manufactured fertilizers onto trucks. Often I rode with the truck drivers all the way to the farms in the San Joaquin, Imperial, and Coachella valleys of California. On several occasions I worked on fertilizer demonstration plots in the potato-growing regions around Bakersfield and the melon-growing districts around Brawley and El Centro. It was fascinating work. I studied agriculture at California State Polytechnic University in San Luis Obispo from 1948 to 1952, majoring in truck crop production and irrigation. Three years of graduate study at Boston University included cross-cultural studies, theology and social ethics. My wife and I went on to Brussels, Belgium, where, at the Ecole Coloniale, we studied French which we would need for agricultural mission service. It was out of this background that I began a lifelong career in agriculture.

My first assignment took me to the Katanga Province of the Belgian Congo. (The colony became independent in 1960 and is now the province of Shaba in the Republic of Zaire.) There I was to work with indigenous village folk (members of the Chokwe and Lunda tribes) to contribute to the development of their agriculture and thereby improve nutrition and overcome rural poverty.

Along the way, during the years of doctoral studies (1966–69), I worked as an agricultural program designer and trainer for the U.S. Peace Corps in French-speaking West Africa. From 1968 through to 1973, with a little overlap in program responsibilities, I was the agricultural programs officer for the World Division of the General Board of Global Ministries of the United Methodist Church. In that position I took part in an ecumenical consortium with agricultural development workers in Southeast Asia, some of the Pacific and Caribbean Islands, Bolivia, most of Africa south of the Sahara, and India and Pakistan. Later on (1974-75) I worked in Niger on a unique rice-growing project—one part of a larger effort to overcome famine conditions created during prolonged drought across the western Sahel from 1969 to 1973. This project received funding from the Lilly Endowment.

But it was in 1959, in the Belgian Congo, that I began to ponder the history of worldwide agriculture and to see its legacies in our time. One August morning I and one of my Belgian friends who worked in agricultural development for the colonial office flew in a small aircraft toward the city of Lubumbashi. I still recall our conversation as we looked out over the vast bush-lands of the Katanga. I asked him how he and his fellow workers within the colonial government's agricultural office understood agricultural development. His reply was clear: the purpose of agricultural development was "to bring another region into the world economy." In this case he referred to the cotton, groundnuts, maize, and cattle export program. He explained how these crops related to Belgian colonial interests. My own assignment was quite different: to address the problem of malnutrition and rural poverty through agricultural development. For my Belgian friend, agriculture was supposed to contribute to the colonial economy. I thought that agriculture was for a local community's welfare. And so I began to ask myself whether agriculture should serve the economy or the economy serve agriculture.

Within the two tribal cultures in which we worked, it was impossible to separate agriculture from human values and social patterns. We were living within an "agri-culture." Customs and tribal traditions sprang from the land. Personal identity was bound to the land and its resources—trees, grasslands, rivers, streams, and fish, animals, and birds. Social status was measured by one's contribution to the community, including the maintenance of soil fertility and forest and water resources. The wisdom and history of the ancestors was remembered and the generations yet unborn were considered. Ritual, taboo, and tribal initiation instilled values and traditions that maintained harmony between people and land. As I learned later on, after overcoming my own cultural orientation, ecological understandings in the Katanga Province were quite sophisticated in terms of what we now call agroecology.[1] Poverty and malnutrition were not the fault of tribal culture, but were wrought by slavery and colonial export cropping systems (in the Katanga this included cotton, groundnuts, and maize). Tribal societies were organized (often by force) to serve the colonial power, be it Belgium, France, Portugal, Italy, Germany, Great Britain or the Netherlands.

The present world food crisis, including the U.S. farm crisis, cannot be understood apart from colonial history. In the closing decades of the 20th century, we still suffer the political, social, economic, and agricultural effects of the slave trade and colonial occupation.

We still cling to the notion (a false one) that wealth generated by the economy and agriculture's contribution to it "trickles down" to benefit rural communities. Our present farm crisis (everywhere in the world) is a "fallout" of this colonial history. The sooner we recognize this fact, the more effective will be our efforts to overcome the farm crisis. But to repeat the point already made, we know the history behind the crisis. It is a history of agriculture designed to serve the powerful at the expense of the powerless. Colonial agriculture was not designed for the welfare of rural people nor for the health of the land, as we shall see in a subsequent chapter. Industrial export agriculture was developed long before Western civilization thought about ecology. Many African societies knew and practiced earth caring, but because slave trade uprooted people and destroyed their cultures, and colonialism further exploited them, much of native wisdom and memory have been lost.

What Agricultural History Teaches

In his booklet *The Conquest of the Land through Seven Thousand Years,* W. C. Lowdermilk traced the long history of agriculture that led to the dust bowl of the late 1930s in the United States.[2] Lowdermilk was the first secretary of the U.S. Soil Conservation Service, an organization that has done much to promote care of the soil. Early in his tenure (1938) he took a leave of absence to travel to Western and Southern Europe, North Africa, and the subcontinent of Asia and China. He wanted to learn what agricultural losses these regions suffered throughout history and to discover how each country addressed the problems of agricultural resource loss. He hoped to gain insights that would be useful in overcoming the tragic loss suffered during the depression and the dust bowl years. Lowdermilk wrote about the "graveyard of empires," pointing out that history was written on the land as civilizations arose and moved eastward to China and westward through Europe and across the Atlantic to the Americas. Throughout history people have changed and often destroyed the land.

For example, 7,000 years ago there were fertile plains in Mesopotamia and the Nile Valley. But the land was heavily grazed as shepherds drove their flocks from Persia to the

Mediterranean Sea. Irrigation systems disrupted normal water cycles, and wandering armies foraged across the land. The record is on the ruined walls of Nebuchadnezzar's Babylon. At least 11 empires rose and fell in this once-fertile land. The Israelites were finally taken captive and made to dig silt out of the canals until foreign invaders brought the maintenance work to an abortive end. At one time, some 25 million people lived on these lands, but now, in the modern state of Iraq, from three to four million people remain. A little further to the west, Israelite flocks and herds overgrazed the Sinai Peninsula and the Trans-Jordan plateau. So it went on the highlands of Judea and Syria as well. Over 4,500 years ago the lush forests of Lebanon were destroyed. During the Roman conquests, the "grain belt" of North Africa gave way to deserts, to dust bowls. The process reached island states of the Mediterranean and even the southern slopes of the European Alps.

A modern example of resource destruction is found in Egypt, where, in the mid-1950s, the Aswan dam was built to store water for irrigation and generate electrical power. But foresight was lacking, and losses are much greater than gains. No longer does the Nile annually flood the agricultural river plains of the delta regions with a thin deposit of regenerating, fertile sediments. The dam brought centuries of this life-giving cycle to an end. Now great quantities of chemical fertilizers are imported to maintain soil fertility for agricultural production. With the ending of annual flooding, Shistomiasis has spread. This is a disease in which the larva of a parasite breed in the slow-moving water of the irrigation canals. They enter through body cavities, penetrating internal organs and causing massive hemorrhaging. Shistomiasis is the number one disease that infects people who work the farms and maintain the irrigation systems. Egypt's agriculture is further aggravated by the buildup of salt in the surface soils, due to high temperatures and scant rainfall. Further downstream where the Nile meets with the sea, the delta lands erode at an alarming rate since there is no more sedimentation from the natural flow of the river. Those sediments, from as far south as Uganda, Sudan, Kenya, and Tanzania, are caught behind the dam. Finally, since natural nutrient flow has been interrupted, the whole food chain of the fisheries of the Mediterranean Sea has been broken. Only a fraction of the sardine and tuna industry remains.

The problem of agricultural loss, however, is by no means confined to the Middle East and the Mediterranean. Far to the east, for example, China had reduced its forests to barren hillsides so that early in our century the rice-growing plains of the great Yellow River were continually silted and flooded, resulting in famine of a magnitude unknown in the history of civilization. Further chapters of this history were later written in the rural poverty of Appalachia and the dust bowl of the North American plains, and are being written today in the ever-growing concentrations of salts in the irrigated soils of the arid West.

After Lowdermilk completed this global investigation, he asked: "If Moses had foreseen what was to become of the promised land, would he have written an eleventh commandment to establish humanity's relation with the land?" He composed the following:

> Thou shalt inherit the Holy Earth as a faithful steward, conserving its resources and productivity from generation to generation. Thou shalt safeguard thy fields from soil erosion, thy living waters from drying up, thy forests from desolation, and protect thy hills from over-grazing by thy herds, that thy descendants may have abundance forever. If any shall fail in this stewardship of the land thy fields shall become sterile, stony ground and wasting gullies, and thy descendants shall decrease and live in poverty or perish from off the face of the earth.[3]

As noted in the first chapter, Donald J. Hughes, in *Ecology of Ancient Civilizations,* also searched the past for reasons for human destruction of the environment. He studied the records of ancient Mediterranean peoples, in particular the Greeks and Romans.

Although Greek culture included admiration for natural beauty, the ancient Greeks had a destructive effect upon the natural environment. They significantly reduced the forests between 600 and 200 B.C. Goats were allowed (as they still are today) to eat every struggling sapling, thereby deforesting countless square miles. The result, as Plato lamented, was the drying of the springs and streams of Attica and the silting of the marshlands and estuaries. Despite prophetic warnings, lumber and firewood were imported to the cities as though the supply of trees would last forever. Overgrazing continued until the land was exhausted. Domestic and export agriculture gradually failed, as did food supplies for the city-states.

The Romans loved their native soil and called it *mater terra,* "mother earth." But whether the earth provided for other people in other nations was no concern of theirs. Rome conquered the earth of the Mediterranean world, including the plants and animals. Nature in Rome's empire was dealt with in strictly utilitarian ways, as one can see from the Roman rituals, designed to control the natural environment and to meet human needs. The gathering of wood for fuel and timber devastated landscapes. Species of great animals were hunted to extinction because the army needed food, crowds at the coliseum needed entertainment, and farmers wanted to raise their livestock without natural predators. The hunting of game animals and birds and the rental of fishing rights accelerated the decline of natural life which reached even to the island communities of Europe's Atlantic coast. Hughes concluded his study of these ancient civilizations from an ecological perspective by saying, "Thus, ancient history is a warning and a challenge to our attitudes and abilities to understand our technological competence and our willingness to make far-reaching decisions."[4] This challenge has yet to be met, as a survey of North American ecological history will show.

In the 1600s, Western European colonial agriculture came to the North American continent. In the southern colonies this meant growing tobacco (and later cotton) for export. The Danes occupied the Caribbean Islands and produced sugar there. Slaves were imported for both the Caribbean and the southern plantations. The "golden triangle" of American trade included rum, guns, and slaves.[5] Agriculture in North America as in the (former) Belgian Congo, was organized to "bring another land area into the economy of western Europe."[6] Refugees from poverty, hunger, and political and religious persecution migrated to the woodlands of the upper Midwest and onto the sodded grasslands of the high plains. Native American people were almost exterminated along with the buffalo.[7] Seeds from northern Europe and red wheat from the Ukraine were introduced.[8] When the native sod was plowed under to make room for wheat, one of the most biologically productive ecosystems on the face of the earth was destroyed. Annual cropping systems prevailed until the "dirty thirties," the dust bowl years.[9]

In 1914 Western Europe faced famine as a consequence of war. The high plains of North America were farmed to produce food for a war-ravaged Europe. Then in the 1940s, the Second World War disrupted worldwide agriculture, and grain had to be grown to win the war. Whether or not grain was ecologically appropriate was not the question. Agriculture was once again designed to meet the needs of national economies, rather than the needs of the land and its farming communities.

It was during World War II (with world populations doubling every thirty years or so) that demographers began to recognize the threat of global famine.[10] Not only was Europe a place of devastation and hunger, but there was malnutrition in Africa, Asia, and Latin America. The colonies of Europe were organized to produce and export minerals, timber, and agricultural products such as coffee, tea, cocoa, tobacco, cotton, hides, sugar, rubber, palm oil, spices, and copra. Domestic food systems were inadequate. Out of the prospect of widespread famine and food deficits in the colonies, the Green Revolution was born. The basic goal of the Green Revolution was to develop self-reliant food production in tropical countries making the transition from colonial to self-determined agriculture. Instead of growing luxury crops (such as coffee and cocoa) for export, nations would grow food for their own people. High-yielding varieties of food grains (wheat, rice, maize) were developed to "buy time." The world's people could be fed for the next 30 years while former colonies became independent states and developed self-reliance in food production.

Certain aspects of the Green Revolution were successful, such as marked increase in crop yields. But the basic goals of self-determination and self-reliance for developing countries have been largely forgotten. We still talk in general terms of international trade advantages.

Today agriculture in the temperate world is almost totally dependent on technologies that require petroleum-based fertilizers and pesticides. We assume that this is the only way to feed a hungry world. A global food system is dependent on one of the earth's most rapidly dwindling resources: oil and natural gas. Most oil and gas will be depleted by the middle of the 21st century, in 50 to 60 years. Meanwhile in the tropical countries, food shortages continue with their grim toll of malnutrition and famine. The agricultural agenda must be a world free from hunger and environmental deterioration.

The development of the high-yielding varieties (HYVs) of rice, wheat, and maize is very significant, but this alone cannot solve the problems of hunger and environmental stress. Agricultural history teaches that in every nation agriculture has become colonialized. That is to say, farmers and rural communities have very little to say about what is to be planted, where it is to be sold, or what finally happens to the land and its people. Nations have surprisingly poor ability to design agriculture to benefit their own people and to enhance their own landscape. The international super-market determines the fate of agriculture. Agriculture has been colonialized everywhere by this market. The result is desertification (dust bowls) on all six continents as well as upon island communities of the Caribbean and Pacific.

In 1969 prolonged drought hit the western Sahel of Africa. By 1974 (as well as in prior years) political conditions in Sahelian West Africa prevented cattle herders from moving their stock as the natural conditions of the rangelands re-quired. This resulted in environmental abuse and finally in famine. At that time there were 28 days of grain in the global food grain reserve. As nations came together for the historic Rome Food Conference, each one presented its strategy for response. The U.S. delegation, led by Secretary of Agriculture Earl Butz, committed the U.S. to a policy of full grain pro-duction. Consequently, 40 million acres of the high plains (placed into the soil bank following the dirty thirties) were opened once again. The new cause: "Food to Feed the World." One wonders if this really meant that there was money to be made once again in the world market. A more enlightened policy would have called for a U.S. commitment to cooperate with other nations in international agricultural development assistance to enable the former colonial world to make the transition from export cropping systems to food self-reliance.

But this did not happen. The international grain market was strong. Grain prices went up. Profits were made up to the mid-1970s. Farmland prices inflated. "Get big or get out" was the slogan and advice from the U.S. Department of Ag-riculture. Following trusted advice from local bankers, the USDA, and market trends, many farmers in the grain states did expand.

Then four things took place. First, because of heavy federal borrowing for a defense buildup, interest rates on crop loans went from 6 percent to more than 18 percent. Second, crude oil prices went from $3 a barrel to $30. Third, Western Europe, Canada, Australia, Thailand, India, and Pak-istan went for full production to compete in the world grain market. This contributed to a grain glut. Finally, the developing nations could no longer borrow funds to purchase needed food grains. The market became flooded; grain prices collapsed and the whole farm economy across all of the grain exporting nations was shaken. Farm foreclosures multiplied, making the farm crisis a national tragedy.

The following story, reported in the *Boston Globe,* September 6, 1987, is but one illustration.

Congressional Representative Richard A. Gephart of Missouri, while on the campaign trail, met with a group of 15 farmers and their spouses in Mount Pleasant, Iowa. Gephardt described these folks as "the cream of the crop, and pillars of the community." One man stood up, and even before he spoke, his wife started to cry. Then the farmer said:

"My son was going to Iowa State. He was in his first year and he came home at Christmas and said: 'Look dad, I don't want to go to school, I just want to be with you. I want to farm.'" This was in 1979.

So they bought another plot to go with the family farm and farmed that. By 1984, they were in trouble with that and they lost it. By '85 they lost the family farm. It had been in the family for three generations.

And he said [that] on the way back from the bank after they had learned they were going to lose it, his son turned to him and said: "Dad, I wish I hadn't come home because … my insisting on coming home and farming is what got you in trouble."

And later in the day the farmer couldn't find him. He went out and opened the barn door and there he was [his son], hanging from the highest rafter.[11]

Gephardt recalled that tears streamed down every face. They all knew each other very well. These neighbors had helped raise this lost son. They knew that this was the first time the farmer had told the story in public. The farmer concluded by saying, "I came here because I think you can do something. We want action. We don't want words. I'm sick of words. I came here because we need hope."[12]

Gephardt reported that he was "knocked out." He said he was dealing with the hardest working people in the world. "They are going under and they are getting cut to pieces."

This story is almost too sacred to use in this book, or any book. Yet it talks about how so many of us feel. This tragic story is not new. It is not unique to the U.S. farm crisis. Too many people can describe similar experiences illustrating what it is like to be colonized, to have agriculture "serve the economy."

When the insight, skill, wisdom, memory, and hope of the farming community is lost (insights developed by generations of families related to the land), desertification takes place both in nature and in the human spirit.[13] This is the lesson that agricultural history has continued to reveal through the centuries. Yet we do not seem to learn the lesson. Agriculture sustains the people, their rural communities, and traditions. Agriculture is far more than an industry for the generation of wealth for special interest groups in the society. History teaches that if the symbiosis or mutual dependence of the land, agriculture, and society is weakened, the society suffers, perhaps even dies.

The crisis in agriculture has been long in the making. In our culture, agriculture is not designed to serve the health of the land and those who till the land. Contemporary political rhetoric is wrong when it claims that lower interest rates and gaining a share of the international market alone will pull us out of the farm crisis. Much more is needed.

The Story of Colonialism

In his opening section of *The Challenge of World Poverty,* Gunnar Myrdal said, "In colonial times and right up to the second world war, the popular as well as the more sophisticated explanations of poverty of the peoples living in what were called "backward regions"—most of them were not "countries"—were, it is now clear in retrospect, plainly apologetic, aimed at relieving the colonial powers and the rich nations generally from moral and political responsibility for the poverty and lack of development of these peoples.[14]

In keeping with these observations, our society needs to take a hard look at colonialism and its legacy. Yet colonialism is seldom studied in public and private schools in the United States. Only those working in some specialized fields at the graduate level encounter this history, which has yet to be adequately researched.[15] The subject is extremely important for the problem this book addresses.

Once again, I am grateful that I had the opportunity to experience Western colonialism when I prepared for agricultural missionary service in the Belgian Congo. I lived through the training program offered by the Belgian colonial office for all missionaries serving in their colony. I experienced the last four years of that colonial occupation and the chaos of the first four years of the Democratic Republic of Zaire. Those were very difficult years indeed.

Colonialism involves the occupation of peoples, their cultures and land, by other peoples and other cultures from other lands. The primary purpose of colonial occupation is to generate wealth and to extend political power. The outsiders who occupy the land want to bring another region into their economy, as stated earlier. Western colonialism, from before the 14th century to the second half of the 20th century, spread across Africa, North and South Asia, Latin America, the Caribbean, and the Pacific Islands. In North America, the fur trade decimated the beaver, otter, mink, and other wildlife along the trade routes, but also in more remote areas. Other parts of the world contributed tea, coffee, cocoa, rubber, hemp, spices, timber, gum, coconut, fishmeal, palm oil, cotton, maize and groundnuts to the European economy. Timber, cattle hides, bananas, and pineapples were added later. The quest for precious gems and minerals, such as gold, silver, iron, copper, boxite, tin, lead, and uranium greatly influenced the settlement or colonization of North and

South America and Africa. Mining, trade, and agriculture in the colonies were not designed for the support of indigenous people. Agriculture was never designed to care for the land. In most cases, all of the wealth went out of the country, resulting in dust bowl conditions and rural poverty. Meanwhile the introduction of modern medicine resulted in rapid population growth, but without adequate food supplies and basic education. This was a formula for disaster. Population in some countries now doubles every 28 to 30 years, a rate unprecedented in human history.[16] Colonialism existed for the benefit of European nations. It was not a program for international development assistance. To the contrary, it destroyed entire peoples whose cultures had once allowed them to relate to the land symbiotically.[17]

Western colonialism came to an end in the early and mid-1960s. During that decade, 85 new nations (former Western colonial possessions) were born. The dramatic shift in the composition of the United Nations since 1960 represents this major change in world political history. Few of these nations, however, were prepared for the change. During the colonial period, people indigenous to the colonial territories (usually males) were trained for the colonial enterprise. There was little or no history, government, political science, sociology, or social ethics taught in the colonies. So when independence arrived, there were no leaders to pick up the reins of government, except those trained by the old occupying powers for service in the colonial police forces. Most of the world's former colonial territories, now independent nations, are governed by third-, fourth-, and fifth-generation military coups d'état. Even now, almost 30 years into the so-called postcolonial era, most developing nations lack the leadership they need to create effective and responsible domestic food systems. Attempts to transplant temperate-zone agriculture into tropical ecosystems have had devastating consequences. Only very recently, following the confident years of the Green Revolution, have there been significant breakthroughs (though nowhere sufficient in terms of need) in the development of technically and socially adequate tropical food systems. Little wonder that so much of the world suffers from hunger, poverty, and environmental stress.

The international business organizations were quick to fill the vacuum created by decolonization. As the Western European colonial ministries withdrew to Europe, international business firms—accountable to no one—moved in. Western industries needed raw materials. So even though the Western nations relinquished political control, the business firms exerted economic control. Neocolonialism was born. Agriculture and other production is still run for the benefit of global markets, generating wealth for corporations and their stockholders. Outsiders continue to control indigenous peoples and occupy their land. In short, agriculture has little to do with our noblesse oblige. The situation is now compounded by the new realities of international debt, debt service, and trade deficits. Colonialism and neocolonialism continue to bear the fruits of poverty, injustice, and desertification.

United States Farm Policy

United States farm policy is also implicated in this history. Seventeen years of effort to develop domestic self-reliance in the former Western colonial territories have given me insight into U.S. agriculture as well. United States agriculture itself has been colonialized. It is designed for the international marketplace, not the rural community, not our national food security, and certainly not the health of the land. It is no wonder that some states are losing 35 percent more topsoil now than at the peak of the dust bowl years. It is no wonder that four million farm families have been displaced since the dust bowl years, with more than 100,000 farm bankruptcies annually. Suicide is on the rise in rural communities. Rural people seem to have little voice in our nation's political process. United States agriculture has been colonialized. Absentee landlords own increasing numbers of farms, many of which are organized into mega-farms run by large corporations. The high price of land and equipment makes it all but impossible for young people to choose farming as a way of life.

In the upper Midwest, a person would have to finance a $640,000 to $700,000 loan in order to start farming a 300- to 400-acre farm. Farmland is about $1,300 per acre at the time this book is being written, used equipment would cost about $100,000, and the beginning farmer would need to pay $150,000 for a house and barn (perhaps a small dairy parlor but no livestock). This is a formidable reality. Such high costs are one of the many reasons that in the upper Midwest, for example, 40 to 50 percent of the farms are now operated by nonfarm owners, perhaps close relatives. Many have had to combine farms. The result is ever-larger acreages in comparison to 20 or 30 years ago.

Today, 4 percent of the U.S. population farms the land, while 96 percent of the society is settled in our great cities and larger towns. Our urbanized and industrialized culture is hardly aware of the farm crisis and is far from comprehending the magnitude of that crisis. Today the great political concerns of the U.S. are with the military/industrial establishment and with urban employment, housing, water and public utilities, drugs, toxic waste disposal, transportation, schools, racial issues, sanitation, retail development, and so on. The needs of the cities determine what happens to the countryside. This is as true for New York, Chicago, Los Angeles, Boston, and New Orleans as it is for New Delhi, Bombay, Calcutta, Beijing, London, Berlin, Paris, Tokyo, Hong Kong, Taipei, Mexico City, Rio de Janeiro, Lagos, Kinshasa, Nairobi, and Cairo. The assumption is that agriculture exists to serve society, and not the other way around. The assumption is that the land and the people who cultivate it can take care of themselves.

At present, U.S. agriculture is supposed to offset the U.S. trade deficits we have incurred through our unceasing demands for more and more goods from the world markets. Our urban society has colonialized U.S. agriculture so that it serves every need except the needs of the land and the people who try to serve the land in a spirit of trusteeship. Colonialism has destroyed every culture that it has ever touched, and in the process the land is also destroyed.

6

Agroecology: A Promising Alternative

Hopeful Signs

There are alternatives to our present agriculture. One hopeful sign is that scientists of biology and earth physics are studying how our biosphere really works. The interdependent relationships of natural systems is a new area of inquiry. A wholistic approach, sensitive to the variations in ecology and population pressures, is becoming more common. The ecology of natural systems has become a new starting point for developing what some call an agroecology (a regenerative agriculture).[1] The goal is to develop an agricultural equilibrium within the natural system.

A second hopeful sign is the reconstruction or at least approximation of the agroecology of precolonial cultures. An attempt is being made to discover and describe the more original landscape, or biotic community, and then to rediscover how people of earlier times related to animal and plant communities.[2] From such an understanding, new goals and guidelines for agriculture can be identified. We cannot turn the population clock back to a quarter of a billion persons. Rather, the land must feed 24 to 26 times as many people as in precolonial times. But better knowledge of earlier biotic communities could provide clues for wiser use of the environment.[3]

An agroecology requires that we preserve and enhance relationships to the land. An agroecology produces food and fiber in ways that enhance the physical and biological environment, at the same time bringing greater dignity to the producing community. A sustainable agriculture restores the land to a semblance of its original form. It allows for the complexity and diversity of a given natural ecosystem. It imitates the original biotic community.

There are three requirements for the survival of any organism, be it a salamander or a human being. They are simple to remember but pose a considerable challenge to our culture. To be regenerative, an agricultural system must function within the following principles:

1. Renewable resources are used within the productive limits of a given environment. Only that which can be replaced is used.
2. All essential nonrenewables are recycled.
3. Waste production does not exceed the environment's capacity to absorb it.

These three biological and physical rules are simple to remember. It is a very different matter, however, to develop a food system capable of supporting the human population within these essential constraints. This is where the challenge lies. There is an additional principle that holds it all together: since all ecosystems are complex, agriculture must also be complex. Agriculture must at the very least permit biological diversity to exist (just the opposite of our present system of biological simplicity). To establish an agroecology we must ask six questions: (1) What was the original ecosystem like before human intervention? (2) How did premodern humanity relate to this environment? (3) What is the environment like today? (4) What changed it? (5) What could be an analogy of those earlier communities? (6) How can we create such an analogy?

We already have some guidelines for meeting this challenge. Regenerative agriculture will be solar and biologically intensive. It will be labor intensive. Farmers will be understood as managers of microbiotic communities that will be smaller than most of today's farms. This new agriculture can be called agroecology. Unlike today's monocropping systems, agroecology will be very complex. Farms will be analogues of earlier ecosystems. Agricultural colleges will be called "Schools of Agroecology," or "Colleges of Biotic Community Management." Farming will be site-specific, harmonizing and enhancing natural diversity of species. Zero tillage (no annual plowing) or perma-culture (production of food, fiber, and fodder crops with perennial grasses and trees) will be the rule, not the exception. Such an agriculture will integrate perennial grasses, trees, and indigenous animal species into the system.[4] It will unlock the genetic potential of more than 20,000 identified edible plants.

Today 85 percent of all food consumed by humanity comes from fourteen plants: wheat, rice, sorghum, millet, corn, barley, bananas, coconuts, cassava, yams, potatoes, soybeans, peas, and table beans.[5] An agroecology will develop the food potential of more plant species. It will also raise indigenous animals such as antelope, bison, elk, and deer. Whether farming is done on the prairie, in woodlands or lowlands, in deserts or tropical forests, or in the oceans (as in aquaculture), each particular environment will have its own specific agricultural methods. In livestock production, farmers will work symbiotically with creatures that have evolved within their ecosystems for tens of thousands of years and have contributed to the health and balance of the plant communities of their natural habitats. New methods for research, food production, and processing will unfold. An agroecology will maintain and enhance the health of the land and those who work upon it. Agribusiness as we define it today will fade just as soon as the price of oil, gasoline, and nitrogen fertilizers skyrockets. Large-scale, capital-intensive agricultural assets will shift to other more profitable places once artificial tax structures that now favor corporate investments in agriculture are modified to reflect a wider justice for people still on the land. When this happens, the search will begin once again for the nearly lost wisdom and memory of those who understood the land as a gift to be held in trusteeship.

An agroecology, or regenerative agriculture, depends upon a secure, numerous, and highly skilled national community of farmers who are well supported by a sound technology, economics, and secure rural society.

A Strategy for Making the Transition

I am indebted to Niel Sampson's work *Farmland or Wasteland*[6] for the following strategy. The first step is crisis awareness building. Within Western society, between 90 and 97 percent of national populations now live in urban and industrial areas. Most of the colleges and universities are urban. Even the religious communities are part of this urban and industrial ethos. Within the non-Western world, although the majority of people live in rural areas, the urban and industrial mindset dominates social policy. The centers of political power are in the great cities, such as Bombay, Calcutta, New York, Washington D.C., Bangkok, Rome, Tokyo, Frankfurt, Dakar, Nairobi, Abdjan, Mexico City, and La Paz.[7] The voice of rural people is not being heard. Insights about rural community and agricultural needs are not given equal time. Those who speak for rural concerns lack political and economic influence. The story about the family tragedy in Iowa, with the farm community's plea for help, is but one illustration. The first task is to make people aware of the crisis in agriculture. This calls for a strong coalition of workers across both the rural and urban sectors. Crisis awareness education requires more teaching about the historical background and the social and environmental dimensions of the crisis. As yet, little has been written on the relationship of the crisis in agriculture to the culture as a whole.

A second step is to work for the preservation of prime agricultural lands. In the radically changed world of the 21st century where only 5 percent (or less) of the earth's surface remains arable for the support of more than six billion persons, the need to preserve prime agricultural land is obvious. Yet, particularly in the United States, such lands continue to be destroyed by urban and industrial sprawl, which is euphemistically called "development." I have observed this rapid transformation during my lifetime as a citizen of the Pacific Southwest. Prime agricultural land conversion continues to gain in momentum as more people migrate to the sun belt, searching for warmer climate and better employment. However, almost everywhere across the 50 states, one sees the loss of prime agricultural landscapes. We act as though our supply of farmland were unlimited. This is of course a myth. The day will come when we realize that cornfields are more necessary for human survival than shopping malls and suburban development.

The rights of those who own these prime lands cannot be ignored or denied. We need a nationwide policy to place prime agricultural land into permanent farmland trusts. This means that federally funded programs could purchase the rights of those in situations where profit-taking from the sale of good farmland is anticipated due to urban and industrial growth. Although the issue is complex, there have been successful models of development rights being purchased and where farmland trusts have been created.[8] Several European nations established such programs long ago.[9] The development rights of the farm owner are paid (higher value based on actual or potential value) with the placement of the farm into perpetuity as farmland. The farm owner retains the land.

The loss of prime agricultural land threatens our nation's food supply. This in turn jeopardizes our national security. The foundation of national security (as well as global security) is secure food systems.[10] The history of civilization clearly teaches this lesson. The immediate costs of securing development rights are great. But far greater will be the costs (if they can ever be met) of recovering (if at all possible) prime agricultural lands that were lost to "development." We need a national commitment to build a federal program to make an inventory of prime agricultural lands, to provide for their preservation, and to safeguard the welfare of future generations and the health of the biosphere. Time is running short. If this problem is not resolved, our national security will weaken as will our ability to contribute to world food needs in times of emergency. We can no longer assume that national food supplies will remain adequate. Likewise, we can no longer assume that the United States will remain a contributing partner in providing for global food security.

Farm assistance and a moratorium on foreclosure is a third part of the strategy for transition to an agroecology.[11] "We want action," said the Iowa farmer in the story told in chapter 5. This is the plea of farm folk all over the world, whether in Iowa, the Philippines, Bolivia, Zimbabwe, Jamaica, or Fiji. Whatever can be done needs to be done now to prevent further foreclosures and the rending of lives, skills, affections, and integrities in rural communities.[12] Food stamps for farmers is not the solution, though it may be needed in the short run. Refinancing of those families crushed by forces beyond their control would cost less than two or three Trident submarines or even a fraction of current military research. The nation must change its perception of national security. Farmers should not be treated like mute pawns of the global marketplace.

Fourth, production controls and price guarantees are needed. An agroecology functions within the limits of the ecosystem to regenerate or replace what is taken from it. We still do not know how much U.S. agriculture will have to reduce its production to be within the limits of its regenerative capacity. Perhaps a 50 percent reduction in food and feed grain production will be required. In a world that is hungry and still dependent on grain from the temperate zones of the world, it would be unconscionable to cut back until (1) nations can feed themselves (food exports would not be necessary except in an emergency), and (2) agricultural commodity price structures change so as to cover the costs of agroecology and at the same time guarantee the economic stability of the farming community. This is a task for agricultural economists. But to ignore this task simply prolongs rural injustice and environmental stress at every point. Commitment to working for a world without hunger—a regenerative, just, and self-reliant agriculture for every nation—is needed because every agricultural economy that is based on export production is threatened by the stress put on its land and resources. It is not practical to perpetuate a world agriculture that can no longer serve the needs of earth and its people. Agroecology gives us hope for a world without hunger and, at the same time, the possibility of rehabilitating stressed landscapes in almost every nation. Peace will be achieved only when justice and the integrity of the land coincide.

A fifth part of the strategy for developing a new agriculture involves research. This responsibility lies with the nation's universities and land grant colleges of agriculture,[13] but research priorities throughout the land grant system need redefinition. Our society needs to improve its food quality, to reverse environmental damage, and to free itself from dependence on petrochemicals. The short-term goals of agribusiness must be replaced by the long-term needs of society. In other words, the land grant colleges of agriculture need to recommit themselves to strengthening their communities through research, education, and extension services. Rural sociology must become a central program within the public agricultural establishment. Research must be renewed and expanded to better examine the ecology of the land, agriculture, and society, and to recognize their interdependent relationships. Long-term research should be respected (and funded) so that research workers are not seduced by the promise of quick results and publications. New standards should be set by the research establishments, namely, contributions to the integrity, harmony, and beauty of the biotic community.[14]

American taxpayers contribute more than four billion dollars each year for agricultural research in the land grant colleges. More should be provided to offset contributions from the private sector so that agricultural researchers can work for the public good as impartially as possible. The public must ask for wiser use of state funds. Since many international students study in U.S. agricultural schools, there is potential for overcoming centuries of colonial agriculture through the building of self-reliant and regenerative food systems. There is hope for a new and better global food system. The land grant colleges have much to contribute to a new kind of agriculture.

Finally we must work for a foreign policy that will support and participate in international agricultural development. When I first began my work in Africa in the mid-1950s, the colonial world was coming to an end as new nations were beginning to win their independence. The United States had a good reputation because of the Marshall Plan's aid to Western Europe following the Second World War. Many people in Africa, Southeast Asia, and Latin America hope that they also would receive development assistance now that the war is over. New governments need help to build roads, bridges, and storage facilities for better domestic food systems. They hope to change from colonies that produce tobacco, sugarcane, palm oil, copra, cocoa, tea, coffee, rubber, cotton for export to self-reliant nations who can feed their own people. But, for many reasons (East-West ideological competition, military agendas, international debt service), agricultural development assistance does not meet the global challenge.

At the Rome World Food Conference in 1974, the Western nations committed themselves to the strategy of the United Nations World Food Council. This program has four components: (1) early warning system; (2) world food grain reserve; (3) food aid; and (4) international fund for agricultural development. This is the basis of an enlightened foreign policy for agricultural development assistance to overcome food deficits in more than 100 nations (about 40 nations are now seriously affected by food shortages).

What Kind of Agriculture Is Efficient?

If agroecology is essential for the postmodern world, the implications are clear. Agriculture must be designed to fit within the pattern of natural systems. "Economy" must now be defined in ecological terms. "Efficiency" has to be measured not merely by yields and profits (which do not count the human and environmental costs) but by the health of the soil, the diversity of species in the area, and the preservation and enhancement of the natural system. Efficiency also includes the well-being (health and nutrition) of the society and the farmer. This is not to say that the ecosystem is more important than the human community. Rather, in an ecological frame of reference, all of life is valued. Nothing is wasted. Wastefulness is the opposite of efficiency. An agriculture that wastes the soil, the ecosystem, and the rural community is not efficient. True efficiency contributes to the total welfare.

One aspect of efficiency is ensuring that farmers receive a price that covers the cost of production with sufficient profit to guarantee the well-being and dignity of the producing community. This means assuring adequate income for the health, housing, education, recreation, and retirement of the farm family. Inheritance tax structures ought to guarantee smooth transitions of farm property from one generation to the next. Implied in this is the preservation of prime agricultural land in the form of agricultural land trusts so that the farming community is free of the threats of development with inevitable changes in property taxation, assessment, and land speculation.[15] Our society has yet to address these issues in a way that is just and consistent. The rehabilitation, preservation, and enhancement of the rural community is basic to the equation. The argument for the preservation of the family farm goes far beyond the Jeffersonian notion that a democracy is founded upon a landed population. There should be as many "managers" of these natural communities as there are natural communities themselves. Further, if integrity, harmony, and beauty are figured into the equation of efficiency, the integrity, harmony, and beauty of the rural community ought to be a central concern of all research for the development of an agroecology. It all goes together.

Agriculture is the responsibility of the entire society. This is as true for agriculture as it is for national defense. Our urbanized and industrialized society is out of touch with its foundation—agriculture. As Hubert Humphrey said, "The truth is that rural America, with the exception of a few of us who want to take up the burden, is forgotten. I mean most of the people in government come out of the great universities, they get a fine education, and by the time they are through with it, what they know about rural America has been flushed out, and they come back with an entirely different set of values and thinking."[16]

The French Experiment

The Republic of France provides us all with a very significant example of agroecology at work. During the early 1950s French agriculture faced three critical problems: fragmented land holdings in the southern provinces, a growing agribusiness with large land holdings in the north, and the inability of young adults to enter farming. In the south, productivity was low, soil loss was widespread, and the watershed was inadequate. For several centuries, farms had been inherited by family members who, more often than not, were absentee owners. This led to low productivity and neglect as farms became ever smaller. As the value of farmland sank, the small towns and villages also declined. The southern provinces no longer engaged in significant food production. Because of large-scale agriculture with centralized marketing in northern France, with its supportive systems of food processing and distribution, the family farm had virtually disappeared.

By 1963 French youth were seeking employment beyond the factories of Paris, Marseilles, and the coastal cities of Normandy. The youth protested that they were unable to enter agriculture as owners/producers. The protest movement resulted in a constitutional amendment that committed the entire republic to the restoration of traditional agriculture. This meant restructuring the nation's agriculture and rehabilitating the rural towns and villages by reestablishing the family farm.

A modest sum of capital was set aside to institute the *Société d'Aménagement Foncier d'Etablissement Rurale* (SAFER), to staff it in agriculturally oriented provinces and empower it to purchase farmland that came onto the open market. The key element of the program was a constitutional amendment that gave SAFER the right of first refusal.[17] This meant that all farmland coming onto the open market had to be considered for purchase by SAFER. If SAFER did not want to purchase the land, it could be sold to any interested buyer. For the first five years there was considerable apprehension about the impact of SAFER on land values. But farmland prices were actually stabilized and improved. Twenty-five years later, about 95 percent of all farmland is sold to SAFER, which in turn holds title to the land for an average of no more than 18 months before it is resold. Capital funds roll over rapidly, with sufficient profit to maintain SAFER's services and to cover any losses incurred.

In each agriculturally based county, SAFER has a staff of anthropologists, rural social scientists, agronomists, surveyors, climatologists, topographers, and specialists in crops, animal husbandry, agricultural economics, and hydrology. In the southern provinces small, fragmented farms are purchased and traded so that they may be consolidated into larger and more productive units. In the north, larger acreages (hectares) are purchased and broken down into family-size units.

In the process of developing family-size units, SAFER's staff appraises the holdings with respect to soil and soil fertility, topography, rainfall and snowfall, existing market infrastructure, and historical and potential recreational settings. If there is a building of historical significance or some potential for the national network of recreational facilities, roadways and fence lines are altered so that there can be multiple uses of a given landscape. Eventually, new farms emerge out of the consolidation process, with farmlands organized according to soil type, climate, and topography. SAFER designates (and sometimes plants) steep slopes as permanent woodlands for the eventual harvest of commercial lumber. Lesser slopes are designated orchard crops and permanent pasture for livestock. Bottomlands of sufficient fertility are designed for annual cropping. After SAFER's improvements, the new farm, economically viable as a family-operated unit, is reevaluated and put back onto the open market. Its size and value are calculated according to its estimated economic viability. Mortgage rates reflect productivity. If the farm is located in a low rainfall area and has lower soil fertility, then not only the price but the terms of the mortgage reflect these natural resource limitations.

Who can purchase these newly constituted farms? To qualify for farm purchase, one must have a certificate to farm. This involves a farming apprenticeship and formal education in agriculture, whether a university degree in agriculture or a technical certificate. Years of successful apprenticeship, recognized by federal authorities, can substitute for formal agricultural education. No one is disqualified due to a lack of formal agricultural education. The point: farm land purchase by speculators is prohibited. In addition, the owner must be the operator and must live continuously on the farm. There are no hobby farms. A new farm owner-operator is required to develop the enterprise within the defined limitations of topography and climate as established by SAFER during the process of consolidation and engineering. There are of course established processes for modification of the farm when better ways are determined.

Productivity is also determined by the market infrastructure of the province in which the farm is located. There is a national effort to decentralize markets in order to stabilize the national food supply, to meet farmers' needs for off-season employment, to provide employment for spouses and working-age farm youth, and to enhance the general economic health of the rural towns and villages.

Things have come alive in rural France! Young people are going into farming. Rural communities offer quality living with amenities of every sort, and rural communities once again offer a good place to raise children. One merely goes to town on market day to see an abundant way of life. The quality of food (from bread to vegetables and fruits, poultry, pork, beef, cheese, butter, and wine) is unsurpassed anywhere in the world. It is exciting to visit farmers' homes and listen to the planning in parlor or kitchen for the children who, if they qualify and hold a farming certificate, will inherit the farm and continue its operation. New woodlots, tree-lined roads, and fence rows grace the horizons. People are there to stay. There is a sense of sacredness. At every other crossroad or corner of a field, one finds a religious shrine. I urge the reader to go and see, taste, and smell. Then let the experience contribute to one's imagination about alternative futures for U.S. agriculture.

Certainly there are many problems to be solved. French agriculture is no utopia. As is true in the United States, French agriculture is a costly, highly subsidized affair. But France (like several other Western European nations) has made its agriculture a high priority. Agriculture is committed to revitalizing the rural community and to improving the health of its countryside. Farmland is a cherished national heritage where trusteeship is exercised. French agriculture may not be a final solution, but it does express ideas about what ought to be . . . and it has structures for working toward those ends. It works to approximate, in significant ways, a socially and environmentally just agroecology. Petrochemicals are still widely used, but the transition is being made to a post-petroleum agriculture. I believe they will make it!

One final point. Today there are several thousand farms in France that are open for affordable family summer vacations. On several occasions my family has vacationed in beautifully reconditioned farm buildings adjacent to lovely pastures, streams, reservoirs, and woodlots. The facilities span from rustic to ultramodern. One can shop for food in the local village, or dine in a first-rate village restaurant. We have purchased milk and butter from the farmstead of our host and hostess and, with the farmer's supervision, chased down a young rooster in the farmyard or a duck at the edge of the pond for our dinner. With tackle furnished by the farm, our children have caught fish in the farm reservoir for either breakfast or dinner. Our family's experiences are shared by thousands of vacationers from around the world. Quality summer recreation is often one important part of the farm income. French agriculture is once again becoming a national heritage. A growing urban-rural solidarity is one of the results of this experience of vacationing. I dream of this for my own country and for all nations.

What Structures of Society Can Contribute

To achieve such solidarity between the people and the land, each part of society must make its contribution. For example, the religious community must exercise its prophetic task, proclaiming judgment and hope. It must reexamine the Scriptures for greater insight into human relationships with the creation. The Scriptures also speak to our attitude toward wealth—indeed our definition of wealth and our responsibility to the neighbor.

There is great need to reexamine the opening chapters of Genesis and to understand the original blessing of creation.[18] So much in Western religious tradition deals with sin and redemption. Is the land included in redemption? And is creation over, or does it continue each day? If it continues, how are we involved in God's creative work? Does the work that we do each day preserve and enhance the creation? Does it serve our neighbors? These are questions that a fresh understanding of creation and agriculture might lead us to ask.

Concerns for the health of the land in prophetic thought, the liturgical references to the land in the Psalms, and the codes of conduct with the land in Deuteronomy deserve closer attention. Claus Westermann (*Elements of Old Testament Theology*), Walter Brueggemann (*The Land*), Rolf Knierim (in his paper "Cosmos and History in Israel's Theology"), and Odil Steck (*World and Environment*) provide examples of theological work in this area.[19]

Religious leadership for rural ministry must be trained to meet the needs of rural communities. Such training could include particular emphasis on Old Testament studies of the theology of the land and the covenant, stewardship, discipleship, and dominion ethics in ancient Hebrew culture. Preparation for rural ministry ought to involve work in the history and structure of American agriculture, farm management concerns, environmental ethics, and farm policy. Carefully supervised rural church and on-farm internships need to be designed for a 12- to 18-month period. Seminaries ought to provide opportunity for a continuous flow of guest speakers, lecturers, and visiting scholars who address rural community and agricultural issues.

Thanks to funding from the Kellogg Foundation, agricultural issues are being studied in several major private liberal arts colleges, and in departments of philosophy, political science and government in some state universities.[20] The purpose is to overcome what the Kellogg Foundation calls "agricultural illiteracy." Many of today's liberal arts students will be tomorrow's political leaders, involved in public policy-making that will affect agriculture. The Kellogg Foundation recognized long ago that agricultural issues are the whole society's responsibility. I had the privilege of working with this program at Pomona College from 1983 to 1986. Since most of the nation's agricultural colleges have narrowed their focus to production efficiencies, national education needs a place to address questions about agricultural history, structure, and farm policy. Areas that need more research include the environmental impact of agriculture on soil and water, environmental toxicity, atmospherics, species decline, the future of a petrochemical and capital-intensive food system, and the stability of the nation's farming community. At the college and university level and beyond, we have begun to call attention to agriculture. We must examine our agriculture from the perspectives of political science, economics, and public policy. Agricultural history needs to be integrated within departments of history. Dissertations wait to be written on the origins of contemporary agriculture. Agriculture needs to be examined within the context of colonialism.

This information could shed light on how we have developed such a destructive, nonsustainable agriculture. From such studies we may learn what will be required to achieve an agriculture that can rehabilitate, preserve, and enhance the land.

In recent years a good start has been made to establish programs (centers, institutes, curricula) for agricultural studies outside the state agricultural colleges and universities and beyond the narrow focus on production, processing, and distribution. But much more needs to be done at every point in the nation's educational network.

The agricultural crisis is a crisis of culture. Until society can more fully understand its responsibility for safeguarding the health and integrity of its agriculture, the agricultural crisis will not be resolved. Society must serve its agriculture. The French effort suggests that this can be done. We shall continue on our collision course with land exhaustion if agriculture continues to subsidize national affluence (affluence for some, not all of the people). If soil, water, oil, and gas resources, as well as the resources of rural communities and farmers, continue to subsidize a national policy of cheap food, then before too long the whole nation will suffer destruction of its agriculture.

The task of the land grant agricultural colleges situated within our state universities is to broaden programs in agricultural engineering, cereal and vegetable crops, rangeland management, soils, irrigation, orchard crops, plant pathology, entomology, and dairy, poultry, and swine production. But these are not enough. Every agricultural school ought to offer classes in agricultural ethics and rural sociology, the history of agriculture and agricultural marketing and legal structures, agroecology, and international agricultural development assistance. The social and environmental issues must be clearly understood. Some of the nation's agricultural schools have taken significant steps in these directions, but these are only a beginning.

The legal aspects of the farm crisis must also be considered. What if Congress were to pass a moratorium on small-farm foreclosures? What if laws were passed to develop state and federal prime agricultural land trusts, and to fund research for and development of a postmodern agriculture? Concerned citizens could work for controls to govern production, pricing, and the use of water, soil, and animal and plant species.

We should also legislate additional funds for the centers of agricultural research within our college and university systems. An annual budget of about $8 billion to $10 billion would probably be adequate at this time. This is but a small fraction of our annual national budget; it is a modest amount to invest for national food and land security. The prevailing farm policy of cheap food for the nation is based on huge subsidies: soil, water, oil, gas, and bankrupt farmers holding at this time more than a $225 billion debt.

More than four million farms have been lost since the Second World War. After the tragic dust bowl years, the nation gave significant priority to the rehabilitation and care of the

101

land. But in the wake of the Second World War, the wars in Korea and Vietnam, and the relentless ideological conflict between East and West, concern for the health of the land and the people who cultivate it gave way to other national preoccupations. The defense establishment and the general economy of the nation, now experiencing a level of spending unimagined 20 years ago (in spite of the fact that more than 20 million citizens are below the poverty line), have eclipsed social concern for the health of the land and the nation's agriculture. The farm crisis is an early symptom of this larger cultural crisis. The color and texture of the waters of the nation's great rivers give further witness (as do the great rivers of the world's continents) to this crisis.

Caring for the earth is expressed in social structures of trusteeship, in codes, regulations, and laws, as well as investments of time, money, and imagination. Trusteeship has as its basic challenge the conquest of hunger, famine, and desertification. Trusteeship for the earth requires an agriculture that rehabilitates, preserves, and enhances landscapes that have potential for agriculture.[21] The care of the land is everyone's responsibility all of the time. Caring will reverse the destruction of the land before it is too late.

7

Ethics for Agriculture

Agricultural Practices and Planetary Survival

One hundred years ago, new agricultural lands waited to be settled. The chemistry and physical composition of the atmosphere (carbon dioxide, nitrogen, ozone, and dust particles) was relatively unchanged from earlier times. Then the earth was able to recycle most of the wastes produced by human industries. At the beginning of this century, the world was a relatively large place; only a few individuals were able to circumvent its breadth.

Today's agriculture was conceived in a world that is long past, a world that was large and pristine. Can contemporary agriculture serve 21st-century needs? Can it serve a world that is small and dangerously polluted?

We do live in an endangered world. During the second half of this century, humanity has seriously damaged the biosphere. According to the United Nations World Conference on the Human Environment which met in Stockholm in 1972, "the two worlds of humanity—the biosphere of our inheritance, the technosphere of our creation—are out of balance and in deep conflict. This is the hinge of history at which we stand, the door of the future opening onto a crisis more sudden, more global, and more inescapable than any ever encountered by the human species."[1]

Indeed, we do stand before a door opening onto a precarious future. The human threat to existence must be overcome during our children's lifetime. And agriculture is a critical element in the quest for planetary survival.

The future of humanity depends upon its commitment to the idea of the global community. From an ecological point of view, community involves all living things, of which humankind is but one member. Now that our species is so numerous, the need to maintain the health and future of all living things is apparent. The health of the planet is the ultimate measure of responsible community. This commitment to life replaces the narrow commercial and political interests that have predominated until now. We exist and can only exist if the health of all life and the integrity of its many patterns of sustenance is maintained. We are part of the "common unity," a term that indicates that we are one species among many and that our future is bound to the future of all the other life on this planet.

Our uniqueness as humans, however, mandates that we play a caring role. We care because we are grateful for life in community and want to preserve it. Since food, air, and water are essential for all life, agriculture plays an absolutely critical role in the process of caring. The essence of agriculture is that it expresses caring in the full sense of the world.

The Urgent Need of Ethics for Agriculture

How does the society create and maintain an agroecology? The point has been made that agriculture is the responsibility of the entire society, not simply of the rural sector. To create and maintain an agroecology, society urgently needs ethics for agriculture.

Books have been written about sexual ethics, medical ethics, and business ethics. New work is being done in the field of environmental ethics. But, strange as it seems, little has yet been attempted in the specific field of agricultural ethics.[2] We have very little to work with in exploring the simple question, What is good agriculture? It is not possible to envision a postmodern agriculture with any integrity without asking the question of agricultural ethics. Our society desperately needs a substantial agricultural ethic. The work of the Department of Philosophy at Florida State University, with its project of Agriculture and Human Values under the leadership of Richard Haines, may be a first in the field of agricultural ethics.[3]

I submit an outline here for a responsible, postmodern agricultural ethic.[4]

Definition

A responsible agriculture enhances the natural system with which it interacts.

Guidelines

A. Species are preserved
B. The health and fertility of the land increases from generation to generation
C. Beauty and justice in personal and community relationships are experienced
D. Agricultural technologies for the production of food and fiber are self-reliant, sustainable, and regenerative

Goals

A. Achieve beauty by enhancing the relationships of land, agriculture, and society
B. Contribute to the common unity
C. Care for the earth

Strategy

A. Conduct research to develop agroecology
B. Rehabilitate and enhance the rural community

C. Legislate to protect prime farmland for the present and future and to control production and pricing in order to secure the economic and social stability of the farming community and the health of the land

Efficiency

Efficiency is measured by self-reliance on a regenerative basis, enhancement of relationships between the land, society, and the agricultural sector, and by the ability of each nation to feed its own people except in the case of emergency.

Values

A. The health and future of the land
B. Quality of human relationships with the land
C. Justice (social and environmental/species)
D. Integrity in work and relationships

Moral motivation

A. Gratitude
B. Responsibility
C. Human purpose of caring for the earth (having dominion)

This outline may seem abstract, but it is exciting to see how it can work. I have seen farms in England, France, Switzerland, Holland, and Japan that are putting agroecology into practice. It is quite an experience to visit an Amish farm where, because of more organic-based fertilizers, there are more earthworms in the soil and more species of birds that multiply and give life and song back to the land from generation to generation. It is exciting to see increased grass density on a game ranch on a semiarid landscape and at the same time observe the ecological contribution of indigenous animal species. In some places the soil is deeper now than in the past. It need not go the other way. As one farmer said, "If you don't know how to leave it in a better condition than when you took it over, then you have no business farming in the first place!" New life is being generated in rural communities in some places in the world. I have listened to Beethoven in rural Japan. I have seen farm families happily at work because work is an honor, service, and opportunity. The ecology of the water buffalo wallow in Sri Lanka, having existed for more than 1,000 years, further illustrates self-reliance and productivity without the use of petrochemicals. Humanity *has* participated creatively in the care of the land.

But when we talk about a new agriculture, and look for signs of hope, we do so within a very sobering context. Our society has no sound standard from which to judge whether our agriculture is good or bad or something in between. Good agriculture is defined by yields per acre. The research establishment focuses on increased yields and tolerance to salinization, but we must not be deceived by the declarations of new breakthroughs. The agricultural crisis will not be over if we regain our former position within the international marketplace. It will not go away with lower interest rates and reduced costs of production.

Guaranteeing the Health of the Land

The haunting questions lie before us. Can we organize society so that it guarantees the health of the land? Can we care for the earth? Can we build an ethic for agriculture? These three questions are intimately related.

We have lived long enough with the assumption that in order to be successful, agriculture must control and manipulate the natural system. Biological complexity has been replaced by vast monocropping systems. Only by the use of chemicals have these systems been maintained, at least for a relatively short period of time.

Within our agriculture today, the raising of livestock has been specialized. There are beef, dairy, poultry, and swine industries. Animal waste has become a problem to be disposed of rather than a precious resource to be used. Crop, pasture, and livestock integration has become a thing of the past. What were rolling dairy pastures are now planted in soybeans and corn. Frequently dairy cows are permanently tethered in the dairy parlor. Natural systems are so completely altered that nonhuman life (cattle, poultry, etc.) bears little resemblance to its original pattern. The integrity of natural systems is destroyed. But the fragile land has limits. Anyone who works in tropical environments quickly learns what the forest teaches. In the temperate regions of the world we have not yet taken the lesson seriously because our soils are young and deep, climates are more forgiving. But soil deterioration goes on. It is just a matter of time before a great deal of our landscape will look like the barren vistas of Ethiopia, Botswana, and southern Zimbabwe.

The May 21, 1989, the Minneapolis *Star Tribune,* reporting on drought in Kansas, noted: "Drought, persisting in large pockets of the nation's heartland from the Texas Panhandle to the Red River Valley, is forcing hundreds of farmers and ranchers to cattle auctions. . . . And it's setting up millions of acres for an encore of the Dust Bowl days when tons of topsoil took flight."

As I think about hope for a new agriculture, I recall a story I heard long ago. A sailing ship had been at sea for many weeks. By the time the ship reached the east coast of South America, the crew had no drinking water left on board. One morning as they neared the equatorial coast, the crew sighted another ship. As they neared it, they called out for fresh drinking water. The other crew replied, "Take your buckets and dip down where you are!" and sailed away without stopping. The thirsty men were outraged. But one sailor did bring up a bucket of water from the sea. It was fresh water. Was this a miracle? Not at all, for the ship was resting off the Brazilian coast at the mouth of the Amazon River, whose fresh waters press several miles into the sea.

Where is hope for a new agriculture? If we dip down into the waters around us, we will find reason to hope. Some exciting beginnings are already under way.

Thanks to a decade of effort by many individuals and groups across the nation, more people than ever before are aware of ecological issues. This work of public education needs to be continued. Conferences and workshops, special meetings, lectures, and seminars increase public awareness about U.S. agriculture and how it works. Public concern could precipitate significant change in research for a new agriculture.

Meanwhile, basic assumptions about agriculture are being questioned by new developments in philosophy, sociology, theology, social ethics, and in science and industry.[5] This is good news, even though it is often painful to change long-cherished assumptions. But new questions, like new seeds, can germinate and bear good fruit.

New values are changing our culture. The feminist movement, for example, has challenged accepted values. Commitments to values such as nurture, cooperation, nonviolence, community, beauty, and self-determination are taking root.

A new generation asks whether agriculture exists to generate wealth or to care for land and people. Do people farm to balance trade and budget deficits? Or do people farm because they love the land and its people? We may even conclude that the purpose of agriculture is to preserve and enhance the land. Here, the question of human purpose is intimately related to agriculture.

Since the United Nations World Conference on Desertification in 1979, we know that during recorded history, more than half the planet's arable soil base has been lost. Given the present momentum of soil loss from erosion and loss of agricultural land to urban and industrial expansion, by the beginning of the 21st century, only 5 percent of the earth's surface will remain arable. The land is fragile and requires constant care. We have realized that its health cannot be taken for granted. Long ago people like Liberty Hyde Bailey believed that those who accept responsibility for trusteeship of the land must do so with humility and reverence. The farmer represents the society by caring for and exercising dominion over the land. The land has its own integrity, which needs to be respected. The land does not belong to us. It is a part of the life of the planet. We are not free to do with it as we please. Like the sea and the air, it is a common heritage. The health of the land is beginning to be recognized as an ultimate good. In the debate on the farm crisis, we are gaining new clarity about how we are all interdependent. This recognition, nearly lost in our Western culture, is now being restored. This is a further reason for hope.

The modern worldview is mechanistic and utilitarian. Our science and technology are shaped by this view. Only that which is useful to us has value. Relationships are understood in terms of competition and conflict rather than responsibility, cooperation, and support. Modern agriculture is coherent and consistent with such a worldview.

But a postmodern worldview is in the making, thanks in part to the environmental movement with its emphasis on the relationships of all forms of life to the environment. As we have seen, ecology avoids concentrating too long on any one entity in isolation of others. The essence of ecological thought is the health and stability of the whole community.

Our task, then, is to develop our agricultural technology and industry in ways that regenerate the land. The goal of agricultural ethics is to enhance the natural system with which it relates. This enhancing is measured or evaluated in terms of species preservation, the health and fertility of the land from generation to generation, and beauty and justice in personal relationships with the land.

Confronted by the farm crisis, the society has the choice to do as it pleases with the land (and take the consequences) or to shoulder the responsibility of trusteeship as an expression of gratitude. Just as a trustee of a college, university, church, synagogue, or hospital can either exploit the trusteeship for personal gain or build responsibly for the sake of the future, so our society must choose how it will relate to the land.

Trusteeship of the land is expressed by the religious community, centers of higher education, and legislation. In the legislative arena, environmental impact reports are required for almost all so-called development activities. Efforts continue to expand in the area of farmland trusts, and most recently, tropical forest preservation. Through the legal and legislative process, progress has been made in environmental protection. When I began my work in California agriculture during the Second World War, not one of these things was a public concern. Times are changing. We have reason to hope.

The human purpose is to care for the earth. If we do this, hunger and famine need not be repeated. Only if we work with an ecological point of view will we be able to overcome the farm crisis and resolve the conflict between the technosphere of our creation and the biosphere of our inheritance. These things must be accomplished during the generations already born. Only out of gratitude for life and our moment in it will we find the will to work for the health of the land and its people. *TAK!*

NOTES

Preface

1. C. Dean Freudenberger and Paul M. Minus, Jr., *Christian Responsibility in a Hungry World* (Nashville: Abingdon, 1976).
2. *A World Hungry: A Resource on Hunger and Hope,* executive producer Karl Holtsnider, based on the work of C. Dean Freudenberger (Los Angeles: Franciscan Communications Center, 1976), filmstrips.
3. C. Dean Freudenberger, "Managing the Land and Water," in *Farming the Lord's Land: Christian Perspectives on American Agriculture,* ed. Charles P. Lutz (Minneapolis: Augsburg, 1980), 123–45.
4. C. Dean Freudenberger, *Food for Tomorrow?* (Minneapolis: Augsburg, 1984).
5. National Agricultural Lands Study, *Final Report* (Washington, D.C.: National Agricultural Lands Study, 1981).
6. Freudenberger, *Food for Tomorrow?*, 16.

Chapter 1

1. J. Donald Hughes, *Ecology in Ancient Civilizations* (Albuquerque: University of New Mexico Press, 1975). Every serious student should consult this work for developing a good perspective on contemporary agriculture.
2. Ibid., 216.
3. See Rachel Carson, *Silent Spring* (Boston: Houghton Mifflin, 1962); Paul Sagan, *Cosmos* (New York: Random House, 1980); Norman Myers, *The Sinking Ark* (New York: Pergamon Press, 1979); René Dubos, *The Wooing of Earth* (New York: Scribner, 1981); Aldo Leopold, *A Sand County Almanac, and Sketches Here and There* (New York: Oxford University

Press, 1949); W. C. Lowdermilk, *Conquest of the Land through Seven Thousand Years,* slightly rev. ed., Agriculture Information Bulletin, no. 99 (Washington, D.C.: U.S. Dept. of Agriculture, Soil Conservation Service, 1975); Liberty Hyde Bailey, *The Holy Earth* (New York: Scribner's, 1915; reprint, Ithaca, N.Y.: New York State College of Agriculture and Life Sciences, 1980); Richard St. Barbe Baker, "The Skin of the Earth," *Warm Wind,* Spring 1980, 16; Wes Jackson is featured in Evan Eisenberg, "Back to Eden," *Atlantic Monthly,* November 1989, 57–89; *The Global 2000 Report to the President—Entering the Twenty-First Century,* Gerald O. Barney, study director (Washington, D.C.: U.S. Government Printing Office, 1980–81); *State of the World: A Worldwatch Institute Report on Progress toward a Sustainable Society,* Lester R. Brown, project director (New York: Norton, 1984–).

4. Wes Jackson, *New Roots for Agriculture* (San Francisco: Friends of the Earth, 1980), 7.

5. For further development of this problem see Wesley Granberg-Michaelson, *A Worldly Spirituality: The Call to Take Care of the Earth* (San Francisco: Harper & Row, 1984).

6. See Lawrence Busch and William B. Lacy, *Food Security in the United States* (Boulder, Colo.: Westview Press, 1984).

7. National Research Council of the National Academy of Science, Committee on the Role of Alternative Farming Methods in Modern Production Agriculture, *Alternative Agriculture* (Washington, D.C.: National Academy Press, 1989).

8. For general energy figures related to U.S. agriculture, see David and Marcia Pimentel, *Food, Energy and Society,* Resource and Environmental Science Series (New York: John Wiley, 1979).

9. See the opening section of Gunnar Myrdal, *The Challenge of World Poverty: A World Anti-Poverty Program in Outline* (New York: Pantheon Books, 1970), 1–29.

Chapter 2

1. For a more thorough commentary on the ecological point of view, see Odil Hannes Steck, *World and Environment,* Biblical Encounters series (Nashville: Abingdon, 1978).

2. An excellent discussion of this neglect in contemporary theology is developed in Gustaf Wingren, *Creation and Gospel: The New Situation of European Theology,* Toronto Studies in Theology (New York: Edwin Mellen Press, 1979); and H. Paul Santmire, *The Travail of Nature: The Ambiguous Ecological Promise of Christian Theology* (Philadelphia: Fortress, 1985).

3. Matthew Fox, *Original Blessing: A Primer in Creation Spirituality* (Santa Fe: Bear, 1983).

4. For a full coverage of these biospheric issues which occupy our attention, see the Worldwatch Institute's *Worldwatch Paper* (1975–); its monthly journal, *World Watch* (1988–); and its annual *State of the World* (1984–).
5. Dubos.
6. For additional discussion about this linkage, see Wendell Berry, *Home Economics* (San Francisco: North Point Press, 1987).
7. Dubos, 140.
8. Leopold, 217.
9. Dubos, 61.
10. Ibid., 115.
11. For a good discussion of this problem, see Cynthia Pollock, *Decommissioning: Nuclear Power's Missing Link,* Worldwatch Paper, no. 69 (Washington, D.C.: Worldwatch Institute, 1986).

Chapter 3

1. For an excellent discussion of the meaning of paradigm and how worldview shapes contemporary thinking about science and technology, see David Ray Griffin, ed., *Spirituality and Society: Postmodern Visions,* Suny Series in Constructive Postmodern Thought (Albany: State University of New York Press, 1988).
2. E. F. Schumacher in *Good Work* (New York: Harper & Row, 1979) discusses the importance of raising the question of "human purpose" in association with technological designs and professional engagement.
3. See Alfred North Whitehead, *Adventures of Ideas* (New York: Free Press, 1967).
4. Schumacher, *Good Work.*
5. E. F. Schumacher, *Small Is Beautiful: Economics as if People Mattered* (New York: Harper & Row, 1973).
6. Schumacher, *Good Work,* 123.
7. John M. Rich, *Chief Seattle's Unanswered Challenge: Spoken on the Wild Forest Threshold of the City that Bears His Name* (Fairfield, Wash.: Ye Galleon Press, 1977), 34.
8. Ibid., 40–41.
9. Thomas Kuhn, *The Structure of Scientific Revolutions,* 2d ed. (Chicago: University of Chicago Press, 1970).
10. Ibid., 70.
11. See Lawrence Busch and William B. Lacy, *Science, Agriculture, and the Politics of Research,* Rural Studies Series of the Rural Sociological Society, Westview Special Studies in Agriculture, Science and Policy (Boulder, Colo.: Westview Press, 1983), for a thorough analysis of the process that led to the truncation of agricultural research.

12. Kuhn, 67.
13. Ibid., 77.
14. Ibid., 90.
15. Ibid., 91.
16. See reports during 1987-88 in *Alternative Agricultural News* referencing the search for new faculty, available at 9200 Edmonston Rd., Suite 117, Greenbelt, MD 20770.

Chapter 4

1. For background to this statement, see Secretariat of the United Nations Conference on Desertification, ed., *Desertification: Its Causes and Consequences* (New York: Pergamon Press, 1977).
2. John A. Livingstone, *One Cosmic Instant: Man's Fleeting Supremacy* (Boston: Houghton Mifflin, 1973), 33.
3. Walter Ebeling, *The Fruited Plain: The Story of American Agriculture* (Berkeley: University of California Press, 1979), 317.
4. Livingston, 29.
5. Sanat K. Majumder, *The Drama of Man and Nature* (Columbus: Merrill, 1971), 55–56.
6. R. K. O'Nions, P. J. Hamilton, and Norman M. Everson, "The Chemical Evolution of the Earth's Mantle," *Scientific American* 242, 5 (May 1980): 91.
7. Livingston, 69.
8. Majumder, 88.
9. For a full discussion of the necessity of maintaining biological diversity, see Myers, 1–111.
10. Livingston, 36.
11. Majumder, 88.
12. Ibid., 295.
13. Livingston, 43.
14. Ibid., 31.
15. Ibid., 44.
16. Ebeling, 24.
17. Ibid.
18. Stephen H. Schneider with Lynne E. Mesirow, *The Genesis Strategy: Climate and Global Survival* (New York: Plenum Press, 1976), 118.
19. For a fascinating account, see Stephen Moorbath, "The Oldest Rocks and the Growth of Continents," *Scientific American* 236, 3 (March 1977): 92–104.
20. Herman and D. M. Hopkins, "Arctic Oceanic Climate in Late Cenozoic Time," *Science* 209 (August 1980): 57–62.
21. Bailey, 24.
22. Ibid.

23. See Douglas John Hall, *Imaging God: Dominion as Stewardship,* Library of Christian Stewardship (Grand Rapids, Mich.: William B. Eerdmans; New York: Friendship Press for the Commission on Stewardship, National Council of the Churches of Christ in the U.S.A., 1986); Claus Westermann, *Genesis 1–11: A Commentary* (Minneapolis: Augsburg, 1984); and Jürgen Moltmann, *God in Creation: A New Theology of Creation and the Spirit of God* (San Francisco: Harper & Row, 1985).
24. For a full discussion about aristocratic responsibility for the care of the land, see Claus Westermann, *Elements of Old Testament Theology* (Atlanta: John Knox, 1982), 85–117.
25. Dubos.
26. Bailey.
27. Leopold.
28. World Commission on Environment and Development, *Our Common Future* (New York: Oxford University Press, 1978), ix.
29. Ibid.
30. *State of the World: A Worldwatch Institute Report on Progress toward a Sustainable Society,* Lester R. Brown, project director (New York: Norton, 1988), 1.
31. Ibid., 4.

Chapter 5

1. For gaining insights about new approaches, see Paul Richards, *Indigenous Agricultural Revolution: Ecology and Food Production in West Africa* (London: Hutchinson; Boulder, Colo.: Westview Press, 1985); Robert Chambers, *Rural Development: Putting the Last First* (Harlow, England: Longman Scientific & Technical; New York: Wiley, 1983).
2. Lowdermilk.
3. Ibid., 30.
4. Hughes, 156.
5. See Basil Davidson, *Black Mother: The Years of the African Slave Trade* (Boston: Little, Brown, 1961) for a historical description of the "golden triangle."
6. As recounted in chapter 5 of this book, *Global Dust Bowl.*
7. For a description of the original ecosystem of this region, see K. Ross Toole, *The Rape of the Great Plains: Northwest America, Cattle and Coal* (Boston: Little, Brown, 1976).
8. See Ebeling.
9. Donald Worster in *Dust Bowl: The Southern Plains in the 1930's* (New York: Oxford University Press, 1979) gives a full account of the causes leading to the dust bowl period.
10. See William Paddock and Paul Paddock, *Famine, 1975! America's Decision: Who Will Survive?* (Boston: Little, Brown, 1967). The predictions in this work were quite prophetic.

11. John Aloysius Farrell, John Vennochi, and Walter V. Robinson, "Sorrowful Meeting for Rep. Gebhardt," *Boston Globe,* Sunday, 6 September 1987, National/Foreign section, 15.
12. Ibid.
13. Wendell Berry, *The Unsettling of America: Culture and Agriculture* (San Francisco: Sierra Club, 1977). Berry is one of the most descriptive authors in assessing this loss of human spirit, the land, and human skill.
14. Myrdal, 4.
15. For a good orientation to this history, see L. H. Gann and Peter Duignan, eds., *Colonialism in Africa, 1870–1960, vol. 3, Profiles of Change? African Society of Colonial Rule,* ed. V. Turner (London: Cambridge University Press, 1969–1975).
16. One should refer to the work of Barry Commoner, *The Closing Circle: Nature, Man and Technology* (New York: Alfred A. Knopf, 1971); and Paul Ehrlich and Anne H. Ehrlich, *Population, Resources, Environment: Issues in Human Ecology* (San Francisco: W. H. Freeman, 1970).
17. See René Dumont and Marie-France Mottin, *Stranglehold on Africa* (London: A. Deutsch, 1983), for a basic history of colonialism in Africa.

Chapter 6

1. Miguel A. Altieri, *Agroecology: The Scientific Basis of Alternative Agriculture,* Westview Special Studies in Agriculture Science and Policy (Boulder, Colo.: Westview Press, 1987).
2. For an excellent illustration of writing history with an ecological eye, see William Cronon, *Changes in the Land: Indians, Colonists, and the Ecology of New England* (New York: Hill & Wang, 1983).
3. See Kenneth A. Dahlberg, ed., *New Directions for Agriculture and Agricultural Research: Neglected Dimensions and Emerging Alternatives* (Totowa, N.J.: Rowman & Allanheld, 1986), for an exhaustive analysis of how present-day research has become so limited in its considerations and why its approach should be broadened.
4. See Bede Okigbo, "Plant Technology in Today's World and Problems of Continued Widespread Adaptation in Less Developed Countries," in *Proceedings, The World Food Conference 1976,* 27 June—1 July, Iowa State University, Ames, Iowa (Ames: Iowa State University Press, 1977), 464. Okigbo identifies the need to expand research by examining the potential in plant species already known for their nutritional benefits.
5. Ibid.
6. R. Neil Sampson, *Farmland or Wasteland: A Time to Choose* (Emmaus, Penn.: Rodale, 1981), 306–36.
7. Lester R. Brown and Jodi L. Jacobson, *The Future of Urbanization: Facing the Ecological and Economic Constraints,* Worldwatch Paper, no. 77 (Washington, D.C.: Worldwatch Institute, 1987).

8. See the work of the American Farmland Trust and its newsletter, *American Farmland.*

9. See Ann L. Strong, *Land Banking: European Reality, American Prospect,* Johns Hopkins Studies in Urban Affairs (Baltimore: Johns Hopkins University Press, 1979).

10. See Busch and Lacy, *Food Security in the United States.*

11. See Marty Strange, *Family Farming: A New Economic Vision* (Lincoln: University of Nebraska Press, 1988).

12. See Gary Comstock, ed., *Is There a Moral Obligation to Save the Family Farm?* (Ames: Iowa State University Press, 1987).

13. See Dieter T. Hessel, ed., *The Agricultural Mission of Churches and Land-Grant Universities: A Report of an Information Consultation* (Ames: Iowa State University Press, 1980). This work represents a "first" in examining future responsibilities for the agricultural schools.

14. See Don A. Dillman and Daryl J. Hobbs, *Rural Society in the U.S.: Issues for the 1980s,* Rural Studies Series of the Rural Sociological Society (Boulder, Colo.: Westview Press, 1982); and National Research Council of the National Academy of Sciences, Committee on the Role of Alternative Farming Methods in Modern Production Agriculture for works demonstrating the need for expansion of the research agenda.

15. See *American Farmland: The Newsletter of the American Farmland Trust.*

16. Freudenberger, *Food for Tomorrow?,* 71.

17. See Strong, 138–236.

18. See Moltmann.

19. For contemporary biblical exegesis developed from ecological perspectives, see Steck; Westermann; Walter Brueggemann, "Land: Fertility and Justice," in *Theology of the Land,* ed. Bernard F. Evans and Gregory D. Cusack (Collegeville, Minn.: Liturgical Press, 1987), 41–68; Rolf Knierim, "Cosmos and History in Israel's Theology," *Horizons in Biblical Theology* 3 (1981): 59–123.

20. Gordon Douglass, ed., *Agricultural Sustainability in a Changing World Order,* Westview Special Studies in Agriculture, Science and Policy (Boulder, Colo.: Westview Press, 1984).

21. The ethical norm of enhancement of the landscape is exhaustively developed in Holmes Rolston, *Environmental Ethics: Duties to and Values in the Natural World* (Philadelphia: Temple University Press, 1988).

Chapter 7

1. Barbara Ward and René Dubos, *Only One Earth: The Care and Maintenance of a Small Planet* (New York: Norton, 1972), 12.

2. The reader is encouraged to see *Journal of Agricultural Ethics,* 1988–.

3. The reader should consult the journals *Agriculture and Human Values,* 1984–; and *Culture and Agriculture,* 1977–.

4. For background to the outline, see C. Dean Freudenberger, "Value and Ethical Dimensions of Alternative Agricultural Approaches: In Quest of a Regenerative and Just Agriculture," in *New Directions for Agriculture and Agricultural Research: Neglected Dimensions and Emerging Alternatives,* ed. Kenneth A. Dahlberg (Totowa, N.J.: Rowman & Allanheld, 1986), 348–64.

5. Background study for the development of an agricultural ethic may be found in Roderick Frazier Nash, *The Rights of Nature: A History of Environmental Ethics* (Madison: University of Wisconsin Press, 1989).

BIBLIOGRAPHY

Agriculture and Human Values, 1984–.

Alternative Agricultural News, 1987–88.

Altieri, Miguel A. *Agroecology: The Scientific Basis of Alternative Agriculture.* Westview Special Studies in Agriculture Science and Policy. Boulder, Colo.: Westview Press, 1987.

American Farmland: The Newsletter of the American Farmland Trust. 1920 N. Street, N.W., Suite 400, Washington, D.C.: 20036.

Bailey, Liberty Hyde. *The Holy Earth.* New York: Scribner's, 1915; reprint, Ithaca, N.Y.: New York State College of Agriculture and Life Sciences, 1980.

Baker, Richard St. Barbe. "The Skin of the Earth." *Warm Wind,* Spring 1980, 16.

Berry, Wendell. *Home Economics.* San Francisco: North Point Press, 1987.

————. *The Unsettling of America: Culture and Agriculture.* San Francisco: Sierra Club, 1977.

Brown, Lester R., and Jacobson, Jodi L. *The Future of Urbanization: Facing the Ecological and Economic Constraints.* Worldwatch Paper, no. 77. Washington, D.C.: Worldwatch Institute, 1987.

Brueggemann, Walter, "Land: Fertility and Justice." In *Theology of the Land,* ed. Bernard F. Evans and Gregory D. Cusack, 41–68. Collegeville, Minn.: Liturgical Press, 1987.

Busch, Lawrence, and Lacy, William B. *Food Security in the United States.* Boulder, Colo.: Westview Press, 1984.

_____.*Science, Agriculture, and the Politics of Research.* Rural Studies Series of the Rural Sociological Society. Westview Special Studies in Agriculture, Science and Policy. Boulder, Colo.: Westview Press, 1983.

Carson, Rachel. *Silent Spring.* Boston: Houghton Mifflin, 1962.

Chambers, Robert. *Rural Development: Putting the Last First.* Harlow, England: Longman Scientific & Technical; New York: Wiley, 1983.

Commoner, Barry. *The Closing Circle: Nature, Man and Technology.* New York: Alfred A. Knopf, 1971.

Comstock, Gary, ed. *Is There a Moral Obligation to Save the Family Farm?* Ames: Iowa State University Press, 1987.

Cronon, William. *Changes in the Land: Indians, Colonists, and the Ecology of New England.* New York: Hill & Wang, 1983.

Culture and Agriculture, 1977–.

Dahlberg, Kenneth A., ed. *New Directions for Agriculture and Agricultural Research: Neglected Dimensions and Emerging Alternatives.* Totowa, N.J.: Rowman & Allanheld, 1986.

Davidson, Basil. *Black Mother: The Years of the African Slave Trade.* Boston: Little, Brown, 1961.

Dillman, Don A., and Hobbs, Daryl J. *Rural Society in the U.S.: Issues for the 1980s.* Rural Studies Series of the Rural Sociological Society. Boulder, Colo.: Westview Press, 1982.

Douglass, Gordon K., ed. *Agricultural Sustainability in a Changing World Order.* Westview Special Studies in Agriculture, Science and Policy. Boulder, Colo.: Westview Press, 1984.

Dubos, René. *The Wooing of Earth.* New York: Scribner, 1980.

Dumont, René, and Mottin, Marie-France. *Stranglehold on Africa.* London: A. Deutsch, 1983.

Ebeling, Walter. *The Fruited Plain: The Story of American Agriculture.* Berkeley: University of California Press, 1979.

Ehrlich, Paul, and Ehrlich, Anne H. *Population, Resources, Environment: Issues in Human Ecology.* San Francisco: W. H. Freeman, 1970.

Eisenberg, Evan. "Back to Eden." *Atlantic* (November 1989): 57–89.

Farrell, John Aloysius, Vennochi, John, and Robinson, Walther. "Sorrowful Meeting for Rep. Gebhardt." *Boston Globe,* 6 September 1987, National/Foreign Section, 15.

Fox, Matthew. *Original Blessing: A Primer in Creation Spirituality.* Santa Fe: Bear, 1983.

Freudenberger, C. Dean. *Food for Tomorrow?* Minneapolis: Augsburg, 1984.

_____. "Managing the Land and Water." In *Farming the Lord's Land: Christian Perspectives on American Agriculture,* ed. Charles P. Lutz, 123–45. Minneapolis: Augsburg, 1980.

_____. "Value and Ethical Dimensions of Alternative Agricultural Approaches: In Quest of a Regenerative and Just Agriculture." In *New Directions for Agriculture and Agricultural Research: Neglected Dimensions and Emerging Alternatives,* ed. Kenneth A. Dahlberg, 348–64. Totowa, N.J.: Rowman & Allanheld, 1986.

Freudenberger, C. Dean, and Minus, Paul M. Jr. *Christian Responsibility in a Hungry World.* Nashville: Abingdon, 1976.

Gann, L. H., and Duignan, Peter, eds. *Colonialism in Africa, 1870– 1960.* Vol. 3, *Profiles of Change: African Society of Colonial Rule,* ed. V. Turner. London: Cambridge University Press, 1969– 75.

The Global 2000 Report to the President—Entering the Twenty- First Century. Gerald O. Barney, study director. Washington, D.C.: U.S. Government Printing Office, 1980–81.

Granberg-Michaelson, Wesley. *A Worldly Spirituality: The Call to Take Care of the Earth.* San Francisco: Harper & Row, 1984.

Griffin, David Ray, ed. *Spirituality and Society: Postmodern Visions.* SUNY Series in Postmodern Thought. Albany: State University of New York Press, 1988.

Hall, Douglas John. *Imaging God: Dominion as Stewardship.* Library of Christian Stewardship. Grand Rapids, Mich.: William B. Eerdmans; New York: Friendship Press for the Commission on Stewardship, National Council of Churches of Christ in the U.S.A., 1986.

Hessel, Dieter T., ed. *The Agricultural Mission of Churches and Land-Grant Universities: A Report of an Information Consultation.* Ames: Iowa State University Press, 1980.

Hopkins, Herman and Hopkins, D. M. "Arctic Oceanic Climate in Late Cenozoic Time." *Science* 209, (August 1980): 557–62.

Hughes, J. Donald. *Ecology in Ancient Civilizations.* Albuquerque: University of New Mexico Press, 1975.

Jackson, Wes. *New Roots for Agriculture.* San Francisco: Friends of the Earth, 1980.

Journal of Agricultural Ethics, 1988–.

Knierim, Rolf. "Cosmos and History in Israel's Theology." *Horizons in Biblical Theology* 3 (1981): 59–123.

Kuhn, Thomas. *The Structure of Scientific Revolutions.* 2d ed. Chicago: University of Chicago Press, 1970.

Leopold, Aldo. *A Sand County Almanac, and Sketches Here and There.* New York: Oxford University Press, 1949.

Livingston, John A. *One Cosmic Instant: Man's Fleeting Supremacy.* Boston: Houghton Mifflin, 1973.

Lowdermilk, W. C. *Conquest of the Land through Seven Thousand Years.* Slightly rev. ed. Agriculture Information Bulletin, no. 99. Washington, D.C.: U.S. Dept. of Agriculture, Soil Conservation Service, 1975.

Majumder, Sanat K. *The Drama of Man and Nature.* Columbus: Merrill, 1971.

Moltmann, Jürgen. *God in Creation: A New Theology of Creation and the Spirit of God.* San Francisco: Harper & Row, 1985.

Moorbath, Stephen. "The Oldest Rocks and the Growth of Continents." *Scientific American* 236, 3 (March 1977): 92–194.

Myers, Norman. *The Sinking Ark.* New York: Pergamon Press, 1979.

Myrdal, Gunnar. *The Challenge of World Poverty: A World Anti-Poverty Program in Outline.* New York: Pantheon Books, 1970.

Nash, Roderick Frazier. *The Rights of Nature: A History of Environmental Ethics.* Madison: University of Wisconsin Press, 1989.

National Agricultural Lands Study. *Final Report.* Washington, D.C.: National Agricultural Lands Study, 1981.

National Research Council of the National Academy of Science. Committee on the Role of Alternative Farming Methods in Modern Production Agriculture. *Alternative Agriculture.* Washington, D.C.: National Academy Press, 1989.

Okigbo, Bede. "Plant Technology in Today's World and Problems of Continued Widespread Adaptation in Less Developed Countries." In *Proceedings, The World Food Conference 1976,* 27 June–1 July, Iowa State University, Ames, Iowa, 451–83. Ames: Iowa State University Press, 1977.

O'Nions, R. K., Hamilton, P. J., and Everson, Norman M. "The Chemical Evolution of the Earth's Mantle." *Scientific American* 242, 5 (May 1980): 90–101.

Paddock, William, and Paddock, Paul. *Famine, 1975! America's Decision: Who Will Survive?* Boston: Little, Brown, 1967.

Pimentel, David, and Pimentel, Marcia. *Food, Energy and Society.* Resource and Environmental Science Series. New York: John Wiley, 1979.

Pollock, Cynthia. *Decommissioning: Nuclear Power's Missing Link.* Worldwatch Paper, no. 69. Washington, D.C.: Worldwatch Institute, 1986.

Rich, John M. *Chief Seattle's Unanswered Challenge: Spoken on the Wild Forest Threshold of the City that Bears His Name.* Fairfield, Wash.: Ye Galleon Press, 1977.

Richards, Paul. *Indigenous Agricultural Revolution: Ecology and Food Production in West Africa.* London: Hutchinson; Boulder, Colo.: Westview Press, 1985.

Rolston, Holmes. *Environmental Ethics: Duties to and Values in the Natural World.* Philadelphia: Temple University Press, 1988.

Sagan, Carl. *Cosmos.* New York: Random House, 1980.

Sampson, R. Neil. *Farmland or Wasteland: A Time to Choose.* Emmaus, Penn.: Rodale, 1981.

Santmire, H. Paul. *The Travail of Nature: The Ambiguous Ecological Promise of Christian Theology.* Philadelphia: Fortress, 1985.

Schneider, Stephen H., and Mesirow, Lynne E. *The Genesis Strategy: Climate and Global Survival.* New York: Plenum Press, 1976.

Schumacher, E. F. *Good Work.* New York: Harper & Row, 1979.

————. *Small Is Beautiful: Economics as if People Mattered.* New York: Harper & Row, 1973.

Secretariat of the United Nations Conference on Desertification, ed. *Desertification: Its Causes and Consequences.* New York: Pergamon Press, 1977.

Steck, Odil Hannes. *World and Environment.* Biblical Encounters Series. Nashville: Abingdon, 1978.

State of the World: A Worldwatch Institute Report on Progress toward a Sustainable Society. Lester R. Brown, project director. New York: Norton, 1984–.

Strange, Marty. *Family Farming: A New Economic Vision.* Lincoln: University of Nebraska Press, 1988.

Strong, Ann L. *Land Banking: European Reality, American Prospect.* Johns Hopkins Studies in Urban Affairs. Baltimore: Johns Hopkins University Press, 1979.

Toole, K. Ross. *The Rape of the Great Plains: Northwest America, Cattle and Coal.* Boston: Little, Brown, 1976.

Ward, Barbara, and Dubos, René. *Only One Earth: The Care and Maintenance of a Small Planet.* New York: Norton, 1972.

Westermann, Claus. *Elements of Old Testament Theology.* Atlanta: John Knox, 1982.

————. *Genesis 1–11: A Commentary.* Minneapolis: Augsburg, 1984.

Whitehead, Alfred North. *Adventures of Ideas.* New York: Free Press, 1967.

Wingren, Gustaf. *Creation and Gospel: The New Situation in European Theology.* Toronto Studies in Theology. New York: Edwin Mellen Press, 1979.

World Commission on Environment and Development. *Our Common Future.* New York: Oxford University Press, 1978.

A World Hungry: A Resource on Hunger and Hope. Executive producer, Karl Holtsnider. Based on the work of C. Dean Freudenberger. Los Angeles: Franciscan Communications Center, 1976. Filmstrips.

World Watch, 1988–.

Worldwatch Paper No. 1– Washington, D.C.: Worldwatch Institute, 1975–.

Worster, Donald. *Dust Bowl: The Southern Plains in the 1930's.* New York: Oxford University Press, 1979.